Agape

Songs of Hope and Reconciliation

Lieder der Hoffnung und Versöhnung

Chants d'Espérance et de Réconciliation

Cantos de Esperanza y Reconciliación

Compiled by Maggie Hamilton and Päivi Jussila

MUSIC DEPARTMENT

OXFORD
UNIVERSITY PRESS

OXFORD
UNIVERSITY PRESS

Great Clarendon Street, Oxford OX2 6DP, England
198 Madison Avenue, New York, NY10016, USA

Oxford University Press is a department of the University of Oxford.
It furthers the University's aim of excellence in research, scholarship,
and education by publishing worldwide in

Oxford New York
Auckland Bangkok Buenos Aires Cape Town Chennai
Dar es Salaam Delhi Hong Kong Istanbul Karachi Kolkata
Kuala Lumpur Madrid Melbourne Mexico City Mumbai Nairobi
São Paulo Shanghai Taipei Tokyo Toronto

Oxford is a registered trade mark of Oxford University Press
in the UK and in certain other countries

1 3 5 7 9 10 8 6 4 2

ISBN 0–19–100023–X

Music and text origination by
Barnes Music Engraving Ltd., East Sussex
Printed in Great Britain on acid-free paper by
Halstan & Co. Ltd., Amersham, Bucks.

Contents

Preface

One of the Church's greatest gifts is its heritage of song. From the days of the earliest Christians, as the gospel has travelled down through the centuries and across continents, poets and musicians have caught the vision of the message and its relevance to their particular context and have expressed their faith through their own words and music.

This legacy is no mere cultural museum. The essence, and saving grace, of the Church has always been that it is a community of people on individual faith journeys who share their insights with others, that all might grow to a deeper and more real relationship with God and understanding of how God moves in our world today. No-one can honestly claim to have a monopoly on God—even the recorded saints of the Church's history struggled to comprehend this strange and mysterious being. By singing songs from the worldwide Church, be they European hymns from the very different context of several centuries ago, or contemporary songs from another continent, we expose ourselves to an encounter with the ultimate truth which may disturb, but which will ultimately lead us to a richer and fuller awareness and experience of God in our daily lives.

As in every generation, our world demonstrates the need for hope and reconciliation on every level, from personal relationships to global justice and peace. This book aims to contribute towards an experience of healing for individuals and societies by sharing songs which communicate a vision of the gospel of all-inclusive love as expressed by Christians around the world. From what we know of it, the early Church met in a spirit of openness and acceptance that encouraged its members to move beyond their current state towards fullness of life. This fellowship of love, or *agape*, is vital to each of us, for it is not by threat, but by invitation, that we are able to move beyond who we are today to become more whole. God has been described as "a beckoning word". Through these songs we are invited and encouraged to make that journey towards wholeness.

Agape was compiled in association with the Lutheran World Federation as a resource for their Assembly in 2003 with the theme "For the Healing of the World". Its reach, however, extends beyond denominational limits. The contents have been compiled from suggestions put forward by an ecumenical team* of musicians, lyricists, and theologians from around the globe and demonstrate that, whatever the language of word and music, the yearning for resurrection in our world is as strong as the faith that dares to believe it is possible.

Maggie Hamilton
London, 2002

*Júlio Cézar Adam, Joy Berg, Marc Chambron, Per Harling, Stephen Larson, Terry MacArthur, Simei Monteiro, Jimson Sanga, Scott Weidler, Dietrich Werner, Mabel Wu, Päivi Jussila (coordinator).

Vorwort

Eines der grössten Geschenke der Kirche ist das Erbe ihrer Lieder. Seit den Tagen der frühesten Christenheit und über die Jahrhunderte hinweg, in denen das Evangelium sich über die Kontinente ausbreitete, haben DichterInnen und MusikerInnen seine Botschaft und dessen Bedeutung für ihren eigenen Kontext aufgegriffen, sowie ihren Glauben durch eigene Worte und Musik ausgedrückt.

Ein solches Erbe ist kein „Kultur-Museum". Das Wesen der Kirche bestand immer darin, dass sie eine Gemeinschaft von Menschen ist, die auf einem eigenen Glaubensweg sind und ihr Verständnis darüber mit anderen teilen. So konnte eine tiefere und wirklichere Beziehung zu Gott wachsen, sowie ein Verstehen dafür, wie Gott in unserer Welt heute handelt.

Kein Mensch kann daher ernsthaft für sich in Anspruch nehmen, ein Monopol auf Gott zu haben. Selbst die Heiligen kämpften darum, dieses fremde und geheimnisvolle Wesen zu verstehen.

Wenn wir Lieder der weltweiten Kirche singen—seien es europäische Kirchenlieder aus der ganz anderen Situation eines vergangenen Jahrhunderts oder Lieder unserer Zeit aus anderen Kontinenten—dann setzen wir uns der Erfahrung der höchsten Wahrheit aus. Sie kann stören. Aber letztlich wird sie uns zu einem reicheren und erfüllteren Bewusstsein der Gotteserfahrung in unserem täglichen Leben führen.

Wie in jeder Generation braucht unsere Welt heute dringend Hoffnung und Versöhnung —auf allen Ebenen, von der Ebene der persönlichen Beziehungen bis hin zu weltweiter Gerechtigkeit und Frieden.

Dieses Buch möchte zu einer Erfahrung von Heilung beitragen, sowohl für einzelne Menschen als auch für Gesellschaften. Es bietet Lieder an, die die Vision des Evangeliums von der allumfassenden Liebe vermitteln, wie sie ChristInnen weltweit weitergeben. Nach dem, was wir wissen, hat sich die frühe Christenheit im Geist der Offenheit und gegenseitigen Akzeptanz getroffen. Das hat ihre einzelnen Glieder bestärkt, über die gegenwärtig prägende Wirklichkeit hinaus zu leben—in Richtung Fülle des Lebens. Diese Geschwisterlichkeit in Liebe—oder auch *agape*—ist lebenswichtig für uns alle. Sie ist nicht zu erzwingen. Sie ist eine offene Einladung, die es uns ermöglicht über das hinaus zu wachsen, was wir heute sind und heil zu werden. Gott ist als „lockendes Wort" beschrieben worden. Durch die Lieder werden wir eingeladen und ermutigt, uns auf den Weg zur Fülle zu machen.

Agape wurde in Zusammenarbeit mit dem Lutherischen Weltbund aus Anlass seiner Vollversammlung 2003 mit dem Thema „Zur Heilung der Welt" zusammengestellt. Das Liederbuch weist jedoch über konfessionelle Grenzen hinaus. Die Liedersammlung basiert auf Vorschlägen eines ökumenischen Teams* von MusikerInnen, LyrikerInnen, und TheologInnen aus der ganzen Welt. Allen Liedern ist, egal in welcher Sprache die Worte und die Musik entstanden sind, die Sehnsucht nach Auferstehung in unserer Welt gemeinsam. Sie ist so stark wie der Glaube, der wagt zu vertrauen: es ist möglich.

Maggie Hamilton
London, 2002

*Júlio Cézar Adam, Joy Berg, Marc Chambron, Per Harling, Stephen Larson, Terry MacArthur, Simei Monteiro, Jimson Sanga, Scott Weidler, Dietrich Werner, Mabel Wu, Päivi Jussila (Koordinatorin).

Avant-propos

L'une des plus grandes richesses de l'Eglise est son patrimoine de chants. Depuis l'époque des premier(ère)s chrétien(ne)s, alors que l'Evangile a traversé les siècles et les continents, poètes et musicien(ne)s ont saisi le sens du message et sa pertinence dans leur situation particulière, et ont exprimé leur foi avec leurs mots à eux, et leur musique.

Cet héritage n'est pas un simple musée culturel. L'essence même de l'Eglise, et sa grâce salvatrice, a toujours montré qu'elle était une communauté d'individus mis en route par leur foi, qui partagent avec les autres leurs découvertes, afin que tous et toutes puissent acquérir une relation plus profonde et plus vraie avec Dieu, et une meilleure compréhension de la façon dont Il se révèle aujourd'hui dans le monde. Personne ne peut vraiment prétendre détenir le monopole sur Dieu, et même celles et ceux qu'on a reconnus pour saints et saintes dans l'histoire de l'Eglise ont eu du mal à comprendre cet être étrange et mystérieux. En chantant les cantiques de l'Eglise universelle, que ce soient des hymnes européens composés il y a plusieurs siècles, ou des chants d'aujourd'hui venant d'autres continents, nous nous exposons à une rencontre avec la vérité absolue qui peut surprendre, mais qui nous conduira finalement à une meilleure compréhension de Dieu, à une plus grande expérience de sa présence dans notre vie quotidienne.

Comme dans chaque génération, notre monde manifeste le besoin d'espérer et de se réconcilier dans bien des domaines, qu'il s'agisse de nos relations personnelles ou de la justice et de la paix mondiales. Ce livre contribue à une expérience de guérison pour les individus et pour les sociétés, en proposant des chants qui nous parlent de l'Evangile et de son amour universel, tels que l'expriment les chrétien(ne)s dans le monde entier. L'Eglise primitive, selon ce que nous en savons, se réunissait dans un esprit d'ouverture qui conduisait ses membres à ne pas rester ce qu'ils étaient, mais à s'ouvrir à une vie plus pleine. Cet amour partagé, ou *agape*, est essentiel pour chacun de nous, car ce n'est pas par la menace, mais en y étant invité(e)s, que nous sommes capables d'aller au-delà de ce que nous sommes aujourd'hui, pour devenir plus complet(ète)s. On a pu parler de Dieu comme d'une «parole qui nous appelle». Par ces cantiques, nous sommes invité(e)s à nous mettre en route vers la plénitude.

Agape a été compilé en partenariat avec la Fédération luthérienne mondiale, en tant qu'instrument pour son Assemblée en 2003 dont le thème sera «Pour guérir le monde». Mais sa portée dépasse les limites confessionnelles. Son contenu est le résultat du travail d'une équipe œcuménique* composée de musicien(ne)s, de parolier(ète)s, et de théologien(ne)s des différentes régions du monde. Il montre bien que, quelles que soient les langues et les mélodies, l'attente de la résurrection dans notre monde est aussi forte que la foi qui ose croire qu'elle est possible.

Maggie Hamilton
Londres, 2002

*Júlio Cézar Adam, Joy Berg, Marc Chambron, Per Harling, Stephen Larson, Terry MacArthur, Simei Monteiro, Jimson Sanga, Scott Weidler, Dietrich Werner, Mabel Wu, Päivi Jussila (coordinatrice).

Prefacio

Uno de los mayores dones de la iglesia es su herencia de canciones. Tal como el evangelio ha viajado a través de los siglos y a través de los continentes a partir de los días de los primeros cristianos y cristianas, así los poetas y los músicos han recogido la visión del mensaje y de su importancia en su determinado contexto y han expresado su fe con sus propias palabras y música.

Esta herencia no es un mero museo cultural. La esencia y gracia salvadora de la iglesia siempre ha radicado en que es una comunidad de gente en peregrinaciones de fe individuales que comparte sus puntos de vista con otros y otras, para que todas las personas puedan crecer hacia una profunda y auténtica relación con Dios, y hacia el entendimiento de cómo Dios se mueve en nuestro mundo de hoy. Nadie puede con honestidad pretender tener el monopolio sobre Dios—incluso los santos conocidos de la historia de la iglesia lucharon para comprender este ser extraño y misterioso. Cantando canciones de la iglesia universal, ya sean himnos europeos de un contexto muy distinto de varios siglos atrás, o canciones contemporáneas de otro continente, nosotros mismos nos exponemos a un encuentro con la última verdad que puede por cierto inquietarnos, pero que por último nos conducirá a un rico y pleno conocimiento y experiencia de Dios en nuestra vida diaria.

Como en cada generación, nuestro mundo demuestra la necesidad de la esperanza y de la reconciliación en cada nivel, desde las relaciones personales hasta la justicia y la paz globales. Este libro intenta contribuir hacia una experiencia de sanación para los individuos y las sociedades, compartiendo las canciones que comunican una visión del evangelio de amor inclusivo, según lo expresado por los cristianos y cristianas de todo el mundo. De lo que sabemos, la iglesia de los primeros tiempos se encontró en un espíritu de apertura y aceptación que animó a sus miembros a moverse más allá de su estado actual hacia la plenitud de la vida. Esta unidad en el amor, o *agape*, es vital para cada uno de nosotros y nosotras, pues no es por amenaza sino por invitación que estamos habilitados a movernos más allá de lo que somos hoy para convertirnos en personas más plenas. Dios ha sido descrito como "una palabra que da señas de invitación". Con estas canciones se nos invita y anima a hacer ese viaje hacia la plenitud.

Agape fue compilado en asociación con la Federación Luterana Mundial como recurso para su Asamblea de 2003 con el tema "Para la Sanación del Mundo". Su alcance, sin embargo, se extiende más allá de los límites denominacionales. El contenido ha sido compilado con sugerencias adelantadas por uno equipo ecuménico* de músicos, líricos, y teólogos alrededor del mundo y demuestra que, cualquiera sea el lenguaje de palabra y música, el deseo para la resurrección en nuestro mundo es tan fuerte como la fe que osa a creer que esto es posible.

Maggie Hamilton
Londres, 2002

*Júlio Cézar Adam, Joy Berg, Marc Chambron, Per Harling, Stephen Larson, Terry MacArthur, Simei Monteiro, Jimson Sanga, Scott Weidler, Dietrich Werner, Mabel Wu, Päivi Jussila (coordinadora).

Notes on the Songs

Languages
- All songs are printed first in the original language, followed in some cases by another language spoken in the region, then translations—if they were available to us—in English, German, French, and Spanish.
- When a text is only a few words, it is printed in the original language only, and given non-singable translations.
- Non-singable translations are given in parentheses.
- Scripts other than Roman have been transliterated.

Music
- As far as possible, we have used original sources. Songs originating in the oral tradition tend to vary according to time and place, so we have published one version only.
- Some songs have been given optional instrumental lines. These are to suggest style, and it is expected that musicians will create their own instrumental parts to match their resources.
- Guitar chords have been given at pitch, so that keyboard players can improvise accompaniments. Keyboard accompaniments have been kept simple, and more accomplished players may wish to elaborate them.
- We have given original metronome markings where these were available. Metronome markings are to be taken as a guide; actual tempi will depend on factors such as acoustic, size of congregation, etc.
- The rhythmic impetus of songs from Africa and Latin America especially means that the singing of them is commonly accompanied by dance and other body movements.

Regional notes
The following notes will enable music leaders to capture the essence of the styles from the different regions.

Africa
- African songs are generally only accompanied by drums and other percussion instruments; South African songs are usually sung *a cappella.*
- Leader lines, and links between verses, tend to be improvised, and vary with each verse. What we have printed is a guide to common practice.

Asia
- Where no accompaniment or chord symbols are given, the most appropriate form of performance is unaccompanied, or with a solo instrument doubling the vocal line. Some songs, such as 'Yarabba ssalami', may be accompanied by oud or guitar playing the melody in repeated notes.

Latin America

- Dance rhythms are indicated and appropriate rhythmic patterns suggested. It is important to follow the rhythm and metronome markings.
- Regional instruments that commonly accompany the songs are listed. In the absence of these instruments, other equivalent instruments may be used. Percussion should be added only to those songs where indicated.
- A useful source of samba patterns is *http://www.channel1.com/users/jfrancis/UBB/patterns.htm.*

Europe and North America

- European and North American hymns, where shown in this book without guitar chords, should normally be sung with keyboard or unaccompanied.

Hinweise zu den Liedern

Sprachen

- Alle Lieder werden zunächst in ihrer Herkunftssprache wiedergegeben, in einigen Fällen folgt eine weitere in der Region gesprochene Sprache, dann folgen—wenn verfügbar—singbare Übersetzungen in Englisch, Deutsch, Französisch, und Spanisch.
- Wenn ein Text nur aus wenigen Worten besteht wird er nur in der Originalsprache wiedergegeben, ergänzt durch nicht-singbare Übersetzungen.
- Nicht-singbare Übersetzungen stehen in Klammern.
- Andere als die lateinischen Schrifttypen werden transkribiert.

Notensatz

- Soweit irgend möglich, haben wir die ursprünglichen musikalischen Quellen und Sätze verwandt. Lieder, die ihren Ursprung in der mündlichen Tradition haben, variieren bisweilen in der Melodie. In diesen Fällen haben wir nur eine Version wiedergeben.
- Bei einigen Liedern haben wir Notensätze für optionale Instrumentalbegleitung hinzugefügt. Diese dienen lediglich dazu, den Stil der Begleitung anzudeuten. Musikalisch Talentierte können ihre eigenen Instrumentalversionen für die Begleitung der Lieder hinzufügen und sie auf ihre Möglichkeiten abstimmen.
- Begleitakkorde für die Gitarre sind in der Normalstimmung hinzugefügt, so dass auch auf dem Keyboard eine entsprechende Begleitung improvisiert werden kann. Begleithinweise für Keyboard sind einfach gehalten, fortgeschrittene Spieler können sie selbstverständlich anspruchsvoller ausarbeiten.
- Metronomangaben sind dem Notensatz hinzugefügt und entsprechen dem Original, wo die Angaben verfügbar waren. Sie sind als Hinweise zu verstehen, die aktuellen Tempi werden sicher auch nach Faktoren wie der akustischen Situation, Grösse der Gemeinde usw. zu gestalten sein.
- Der stärker rhythmisierte Charakter der Lieder aus Afrika und Lateinamerika ist dadurch bedingt, dass das Singen dieser Lieder zumeist durch gemeinsamen Tanz oder andere körperliche Bewegungen begleitet wird.

Hinweise zu den Herkunftsregionen

Die folgenden Hinweise sollen ChorleiterInnen und KirchenmusikerInnen befähigen, den wesentlichen Charakter der musikalischen Stile aus den verschiedenen Regionen genauer zu erfassen.

Afrika

- Lieder aus Afrika werden gewöhnlich nur durch Trommeln oder andere Rhythmusinstrumente begleitet; Lieder aus Südafrika werden zumeist *a cappella* gesungen.
- Melodien für VorsängerInnen und Verbindungsstücke zwischen einzelnen Versen werden häufig improvisiert und verändern sich von Vers zu Vers. Was wir abgedruckt haben ist als Hinweis auf die üblichste Praxis zu verstehen.

Asien

- Wo keine Begleitstimme oder Begleitakkorde angegeben sind, ist die gewöhnlichste Form der musikalischen Gestaltung diejenige ohne Begleitung – oder mit einem Soloinstrument, das die Vokalstimme verdoppelt. Einige Lieder, wie zum Beispiel ‚Yarabba ssalami‘, können durch eine Gitarre oder Oud, die die Melodie wiederholt spielt, begleitet werden.

Lateinamerika

- Angaben zu den Tanz- und Rhythmusstilen sind jeweils den Liedern hinzugefügt. Es ist wichtig, die Rhythmusangaben und Metronomangaben zu beachten.
- Traditionelle Instrumente aus der Region, mit denen die Lieder zumeist begleitet werden, sind genannt. Falls solche Instrumente vor Ort fehlen, können die Lieder mit ähnlichen Instrumenten begleitet werden. Schlagzeug sollte nur bei solchen Lieder hinzugefügt werden, wo es angezeigt ist.
- Eine hilfreiche Übersicht über die verschiedenen Samba-Tanzrhythmen findet sich im Internet unter: *http://www.channel1.com/users/jfrancis/UBB/patterns.htm.*

Europa und Nordamerika

- Lieder aus Europa oder Nordamerika, die ohne Begleitakkorde für Gitarre wiedergegeben werden, sollten normalerweise mit Keyboard oder Orgel-Begleitsatz oder ohne Begleitung gesungen werden.

Remarques sur les chants

Langues

- Tous les chants sont imprimés d'abord dans leur langue d'origine, suivie dans certains cas par une autre langue parlée dans la région, puis par la traduction – quand elle nous était disponible – en anglais, allemand, français et espagnol.
- Quand un texte n'a que quelques mots, il n'est imprimé que dans la langue originale, et l'on donne des traductions non chantables.
- Les traductions non chantables sont entre parenthèses.
- Les écritures non romaines ont été transcrites.

Musique

- Autant que possible, nous avons utilisé les sources d'origine. Les chants venant de la tradition orale ont tendance à changer selon les temps et les lieux, aussi ne publions-nous qu'une seule version.
- Certains chants comportent des lignes instrumentales facultatives qu'on peut ou non choisir. Ceci afin de suggérer un style, avec l'idée que les musicien(ne)s créeront leur propre version instrumentale, selon leurs possibilités.
- Des accords de guitare ont été indiqués pour que les joueurs de clavier puissent improviser les accompagnements. Ceux-ci sont volontairement simples, et des musicien(ne)s accompli(e)s pourront les enrichir.
- Nous avons donné les indications d'origine concernant le métronome quand celles-ci étaient disponibles. Ces indications pour le métronome ne sont qu'un guide, car les véritables tempi dépendront d'éléments tels que l'acoustique, la taille de la communauté, etc.
- L'élan rythmique des chants d'Afrique et d'Amérique Latine indique que ceux-ci sont souvent accompagnés de danses et autres mouvements corporels.

Remarques selon les régions

Les notes qui suivent aideront les dirigeant(e)s à prendre en compte les styles des différentes régions.

Afrique

- Les chants africains sont en général accompagnés seulement par des tambours et autres instruments à percussion; les chants sud-africains sont habituellement chantés *a cappella*.
- Les lignes directrices et les liaisons entre les strophes sont souvent improvisées, et varient avec chaque strophe. Ce que nous avons imprimé est un guide pour chanter en groupe.

Asie

- Lorsque ne sont indiqués ni accompagnement ni accords, il est conseillé d'interpréter sans accompagnement, ou avec un instrument solo doublant la ligne vocale. Certains chants, par exemple, «Yarabba ssalami», peuvent être accompagnés par un oud ou une guitare jouant la mélodie avec des notes répétées.

Amérique Latine

- Les rythmes de danse sont indiqués, et les modèles rythmiques qui conviennent sont suggérés. Il est important de suivre le rythme et les battements du métronome.
- Les instruments régionaux qui devraient normalement accompagner les chants sont indiqués. S'ils ne sont pas disponibles, d'autres instruments équivalents peuvent être utilisés. N'ajouter la percussion à ces chants que lorsque cela est indiqué.
- Une source utile de modèles de sambas est *http://www.channel1.com/users/jfrancis/UBB/patterns.htm.*

Europe et Amérique du Nord

- Les chants européens et nord-américains, quand ils sont dans ce livre sans accords de guitare, doivent être normalement chantés avec clavier, ou sans accompagnement.

Notas referentes a los cantos

Idiomas

- Todas las canciones figuran primero en su idioma original y luego, en algunos casos, en uno u otro idioma hablado en la región y, por fin, si disponíamos de traducciones, en inglés, alemán, francés y español.
- Cuando determinado texto consiste tan sólo de unas pocas palabras, lo hemos publicado solamente en su idioma original, con traducciones que no se prestan a ser cantadas.
- Las traducciones que no se prestan a ser cantadas, figuran entre paréntesis.
- Caligrafías que no son de letra romana han sido transliteradas en forma fonética.

Música

- En lo posible, hemos utilizado las fuentes originales. Canciones que tienen su origen en la tradición oral, pueden ofrecer diferentes variaciones según la época y el lugar de donde provienen, pero nos hemos limitado a tan sólo una versión.
- Algunas canciones figuran con indicaciones instrumentales facultativas para sugerir un estilo, lo que significa que los músicos podrán desarrollar sus propias melodías, según los recursos disponibles.
- Se indican los acordes de guitarra sobre el pentagrama, para que así los intérpretes al teclado puedan improvisar un acompañamiento. Los acordes de acompañamiento son más bien simples, y los intérpretes más experimentados podrán explayarse en el tema musical.
- Las indicaciones metronómicas disponibles figuran en esta obra como guía; el ritmo usado dependerá de factores tales como la acústica, el tamaño de la congregación, etc.
- El ímpetu rítmico de canciones de África y de América Latina sugiere que el canto es comúnmente acompañado de baile u otros movimientos del cuerpo.

Notas regionales

Con las siguientes notas tratamos de ayudar a los directores de música a captar lo esencial de los estilos de las diferentes regiones.

África

- Por lo general, las canciones africanas tan sólo llevan un acompañamiento de tambores y otros instrumentos de percusión; las canciones sudafricanas se suelen cantar sin acompañamiento instrumental.
- Hay cierta tendencia a improvisar la introducción y los enlaces entre las estrofas, que pueden variar para cada una de éstas. Lo que figura en esta publicación, es una guía para lo que comúnmente se practica.

Asia

- Cuando no hemos dado ni acompañamiento ni símbolo para los acordes, la representación más apropiada es sin acompañamiento, o con un instrumento único que duplica la voz humana. Algunas de las canciones, tales como 'Yarabba ssalami', pueden recibir un acompañamiento de oud (una especia de laúd del Oriente Medio) o guitarra repitiendo cada nota varias veces.

América Latina

- Hemos indicado los ritmos de baile y sugerido las formas rítmicas más apropiadas. Es importante seguir el ritmo y las indicaciones metronómicas.
- Hemos enumerado los instrumentos musicales que por lo general sirven de acompañamiento a las canciones publicadas. Si no se disponen de tales instrumentos, pueden usarse otros equivalentes. Tan sólo se deben usar instrumentos de percusión para las canciones en que lo hemos indicado.
- Para ritmos de samba sugerimos consultar el sitio Web: *http://www.channel1.com/users/ jfrancis/UBB/patterns.htm.*

Europa y América del Norte

- Los himnos europeos y norteamericanos que en esta publicación figuran sin acordes de guitarra, se deberán cantar por lo general con instrumentos de teclado o sin acompañamiento.

Agape

1

A ivangeli ya famba

Trad. Xitswa, Mozambique

♩ = 126

LEADER

A i-va-nge-li ya fa-mba, a i-va-nge-li ya fa-mba,

ALL

Ya fa-mba, ya fa-mba,

a i-va-nge-li ya fa-mba,___ yi ya ma-

ya fa-mba, a i-va-nge-li ya fa-mba,___ yi ya ma-

1.
-ti-kwe-nhi hi-nkwa - wo. A i-va- -wo.

2. Fine
Ta-yi tchi-ke-ta le *A-fri-ca,

-ti-kwe-nhi hi-nkwa - wo. -wo.

* insert names of countries or continents, e.g. India, America, Jamaica, UK, etc.

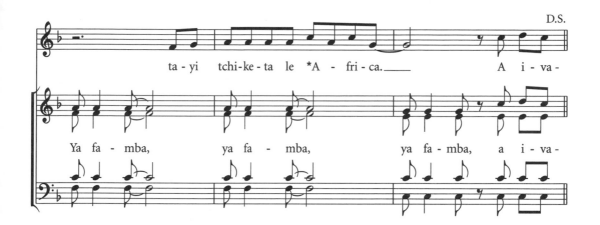

ta - yi   tchi-ke-ta   le   *A - fri - ca.＿＿＿   A   i - va -

Ya   fa - mba,   ya   fa - mba,   ya   fa - mba,   a   i - va -

(The Gospel is going (is going) all over the world.
We will take it to *Africa, India, America, Jamaica, UK, etc.)

(Das Evangelium umrundet (umrundet) die Welt.
Wir bezeugen es in *Afrika, Indien, Amerika, Jamaica, UK, usw.)

(L'Evangile parcourt (parcourt) le monde.
Nous l'emmènerons en *Afrique, en Inde, en Amérique, en Jamaïque, au Royaume-Uni, etc.)

(El Evangelio va (va) por todo el mundo.
Lo llevaremos a *Africa, a India, a América, a Jamaica, al Reino Unido, etc.)

2 | A ti, Señor, te pedimos (Confesión)

Ulises Torres

Trad., Chile

Spanish

1 A ti, Señor, te pedimos
perdón en este momento
por los pecados de acciones,
palabras y pensamientos.

2 De tiempos inmemoriales
que nos hemos separado
de tu comunión bendita:
perdona nuestro pecado.

3 De tiempos inmemoriales
que nos hemos separado
de todos los que tú amas:
perdona nuestro pecado.

4 De tiempos inmemoriales,
en el alma del humano
hay luchas que lo destruyen:
perdona nuestro pecado.

Portuguese

1 A ti, Senhor, te pedimos
o teu perdão, teu alento;
pecamos por nossos atos,
palavras e pensamentos.

2 De tempos já sem memória
vivemos tão separados
da tua mesa bendita:
perdoa nossos pecados.

3 De tempos já sem memória
vivemos tão separados
de todos os que tu amas:
perdoa nossos pecados.

4 De tempos já sem memória
vivemos desesperados
em lutas, ódio e malícia:
perdoa nossos pecados.

tr. Jaci Maraschin

English

1 Have mercy, hear our confession,
O God: blot out our misdoings;
forgive us where selfish motives
have scarred the earth we inhabit.

2 Have mercy, hear our confession,
O God: forgive us the moments
when we have failed with compassion
to heal, encourage, enable.

3 Share out among us your Spirit,
that we may grow in awareness
and thus discern in each other
the human features of Jesus.

4 O God, you send us, forgiven,
into your world every morning
to speak, make peace among people:
each day a hopeful beginning!

tr. Fred Kaan

German

1 Wir bitten dich, Gott, erbarm dich.
Vergib, was wir vor dich bringen,
was uns bedrückt und befremdet,
was Schatten wirft, was uns Angst macht.

2 Wir bitten dich, Gott, vergib uns,
wir haben dich oft verraten.
Wir haben mit dir gebrochen
in Gedanken, Worten und Taten.

3 Wir bitten dich, Gott, vergib uns,
wir leben auf Kosten and'rer.
Weil wir nicht eins mit uns selbst sind,
fehlt uns der Atem zur Liebe.

4 Wir bitten dich, Gott, sei mit uns,
weil wir deine Nähe brauchen.
Wir bitten dich, Gott, bleib mit uns.
Zeig uns die Spuren des Friedens.

tr. Fritz Baltruweit

French

1 Pardon, Seigneur, pour nos fautes,
pour tous les péchés du monde,
pour nos actions malfaisantes,
pour nos paroles blessantes.

2 Depuis trop longtemps, nous sommes
loin de toi et de ta face,
sans communion, sans ta grâce.
Pardon, Seigneur, pour nos fautes.

3 Depuis trop longtemps, nos guerres
font mal à ton cœur de père,
nous qui sommes tes enfants.
Pardon, Seigneur, pour nos fautes.

4 Redonne-nous l'espérance,
et viens guérir nos souffrances,
et tous les péchés du monde.
Pardon, Seigneur, pour nos fautes.

tr. Marc Chambron

GUITAR

A Palavra do Senhor

based on Isaiah 55: 11 and Luke 19: 40

Unknown, South America

Canon

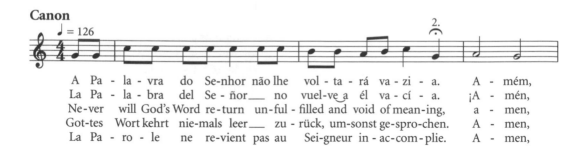

A Pa - la - vra do Se-nhor não lhe vol - ta - rá va - zi - a. A - mém,
La Pa - la - bra del Se - ñor__ no vuel-ve a él va - cí - a. ¡A - mén,
Ne-ver will God's Word re-turn un-ful - filled and void of mean-ing, a - men,
Got-tes Wort kehrt nie-mals leer__ zu - rück, um-sonst ge-spro-chen. A - men,
La Pa - ro - le ne re-vient pas au Sei-gneur in - ac-com - plie. A - men,

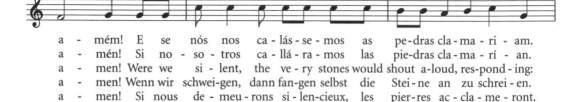

a - mém! E se nós nos ca - lás - se - mos as pe-dras cla-ma - ri - am.
a - mén! Si no - so - tros ca - llá - ra - mos las pie-dras cla-ma - rí - an.
a - men! Were we si - lent, the ve - ry stones would shout a-loud, res-pond - ing:
a - men! Wenn wir schwei-gen, dann fan-gen selbst die Stei-ne an zu schrei - en.
a - men! Si nous de - meu - rons si - len-cieux, les pier-res ac - cla - me - ront.

A - mém, a - mém! A - - - mém, a - mém,
¡A - mén, a - mén! ¡A - - - mén, a - mén,
A - men, a - men! A - - - men, a - men,
A - men, a - men! A - - - men, a - men,
A - men, a - men! A - - - men, a - men,

a - mém, a - mém! A - - mém, a - mém, a - mém!
a - mén, a - mén! ¡A - - mén, a - mén, a - mén!
a - men, a - men! A - - men, a - men, a - men!
a - men, a - men! A - - men, a - men, a - men!
a - men, a - men! A - - men, a - men, a - men!

English tr. Fred Kaan; German tr. Dieter Trautwein; French tr. Béatrice Bengtsson

4 | Agios o Theos

From the Orthodox Liturgy, Greece

Byzantine chant,
Petros Lampadarios

A - gi - os o The - os, a - gi - os i - schy - ros,

a - gi - os a - tha - na - tos, e - le - i - son i - mas.____

(Holy God, Holy Mighty, Holy Immortal: have mercy on us.)

(Heiliger Gott, heiliger Mächtiger, heiliger Unsterblicher, erbarme dich unser.)

(Dieu saint, saint et fort, saint et immortel, aie pitié de nous.)

(Santo Dios, Santo Poderoso, Santo Immortal ¡à ten misericordia de nosotros!)

The key signature indicates that the note A is sung approximately a quarter tone flat.
The G drone should be sung on a neutral vowel.

5

Aleluya Y'in Oluwa

Trad. Yoruba, Nigeria

♩ = 126

A – le – lu – ya Y'in O – lu – wa.____
Al – le – lu – ia, praise the Lord.____
Freu – den – lie – der singt un – serm Gott.____

A – le – lu – ya Y'in O – lu – wa.____
Al – le – lu – ia, praise the Lord.____
Freu – den – lie – der singt un – serm Gott.____

O seun, o seun, o seun, o seun ba – ba
Prais – es, high prais – es, now bring to the Lord.
Tanzt eu – ren, singt eu – ren Lob – preis zu Gott.

A – le – lu – ya Y'in O – lu – wa.____
Al – le – lu – ia, praise the Lord.____
Freu – den – lie – der singt un – serm Gott.____

Yoruba

1 Aleluya Y'in Oluwa.
   Aleluya Y'in Oluwa.
   O seun, o seun, o seun, o seun baba
   Aleluya Y'in Oluwa.

2 E k'orin ayo s'Oluwa.
   E k'orin ayo s'Oluwa.
   ka f'ope fun ka si tun maa jo
   e k'orin ayo s'Oluwa.

3 E k'orin e lu'lu f'Oluwa.
   E k'orin e lu'lu f'Oluwa.
   O seun, o seun, o seun, o seun baba
   e k'orin e lu'lu f'Oluwa.

English

1 Alleluia, praise the Lord.
   Alleluia, praise the Lord.
   Praises, high praises, now bring to the Lord.
   Alleluia, praise the Lord.

2 Songs of joy sing to the Lord.
   Songs of joy sing to the Lord.
   Dancing and singing your praises to God.
   Songs of joy sing to the Lord.

3 Beat the drums; sing out for the Lord.
   Beat the drums; sing out for the Lord.
   Praises, high praises, give thanks to the Lord.
   Beat the drums; sing out for the Lord.
                           tr. Emmanuel Badejo

German

1 Freudenlieder singt unserm Gott.
   Freudenlieder singt unserm Gott.
   Tanzt euren, singt euren Lobpreis zu Gott.
   Freudenlieder singt unserm Gott.

2 Schlagt die Trommel, singt laut zu Gott.
   Schlagt die Trommel, singt laut zu Gott.
   Stampfender Trommelklang preist unsern Gott.
   Schlagt die Trommel, singt laut zu Gott.

3 Halleluja, preist unsern Gott.
   Halleluja, preist unsern Gott.
   Leichtigkeit, Heiterkeit, preist unsern Gott.
   Halleluja, preist unsern Gott.
                           tr. Horst Bracks

6a All people that on earth do dwell

William Kethe
based on Psalm 100

88 88 (LM)
Melody by (or adapt.) Louis Bourgeois
harm. Claude Goudimel

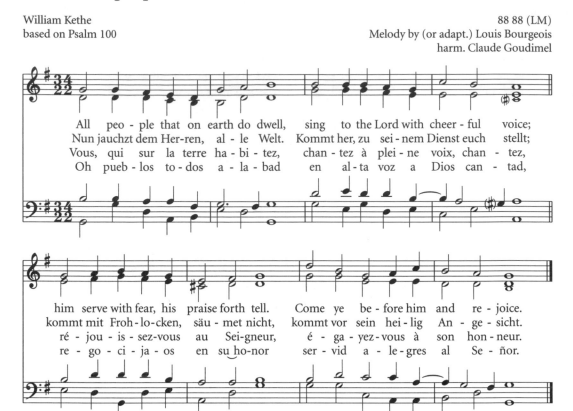

All peo - ple that on earth do dwell, sing to the Lord with cheer - ful voice;
Nun jauchzt dem Her - ren, al - le Welt. Kommt her, zu sei - nem Dienst euch stellt;
Vous, qui sur la terre ha - bi - tez, chan - tez à plei - ne voix, chan - tez,
Oh pueb - los to - dos a - la - bad en al - ta voz a Dios can - tad,

him serve with fear, his praise forth tell. Come ye be - fore him and re - joice.
kommt mit Froh - lo - cken, säu - met nicht, kommt vor sein hei - lig An - ge - sicht.
ré - jou - is - sez - vous au Sei - gneur, é - ga - yez - vous à son hon - neur.
re - go - ci - ja - os en su ho - nor ser - vid a - le - gres al Se - ñor.

English

1 All people that on earth do dwell,
   sing to the Lord with cheerful voice;
   him serve with fear, his praise forth tell.
   Come ye before him and rejoice.

2 The Lord ye know is God indeed;
   without our aid he did us make;
   we are his folk, he doth us feed,
   and for his sheep he doth us take.

3 O enter then his gates with praise;
   approach with joy his courts unto;
   praise, laud, and bless his name always,
   for it is seemly so to do.

4 For why? The Lord our God is good;
   his mercy is for ever sure;
   his truth at all times firmly stood,
   and shall from age to age endure.

German

1 Nun jauchzt dem Herren, alle Welt.
   Kommt her, zu seinem Dienst euch stellt;
   kommt mit Frohlocken, säumet nicht,
   kommt vor sein heilig Angesicht.

2 Erkennt, dass Gott ist unser Herr,
   der uns erschaffen ihm zur Ehr.
   Als guter Hirt ist er bereit,
   zu führen uns auf gute Weid.

3 Die ihr nun wollet bei ihm sein,
   kommt, geht zu seinen Toren ein
   mit Loben durch der Psalmen Klang,
   zu seinem Hause mit Gesang.

4 Er ist voll Güt und Freundlichkeit,
   voll Lieb und Treu zu jeder Zeit.
   Sein Gnad währt immer dort und hier
   und seine Wahrheit für und für.

after Cornelius Becker

French

1 Vous, qui sur la terre habitez,
chantez à pleine voix, chantez,
réjouissez-vous au Seigneur,
égayez-vous à son honneur.

2 Lui seul est notre souverain,
c'est lui qui nous fit de sa main:
nous le peuple qu'il mènera,
le troupeau qu'il rassemblera.

3 Présentez-vous tous devant lui,
dans sa maison dès aujourd'hui;
célébrez son nom glorieux,
exaltez-le jusques aux cieux.

4 Pour toi, Seigneur, que notre amour
se renouvelle chaque jour:
ta bonté, ta fidélité,
demeurent pour l'éternité.

Roger Chapal,
after Théodore de Bèze

Spanish

1 Oh pueblos todos alabad
en alta voz a Dios cantad,
regocijaos en su honor
servid alegres al Señor.

2 El soberano Creador
de nuestra vida es el autor;
el pueblo suyo somos ya,
rebaño que el pastoreará.

3 A su santuario pues entrad,
y vuestras vidas ofrendad;
al nombre augusto dad loor,
que al mundo llena de esplendor.

4 Incomparable es su bondad,
y para siempre su verdad;
de bienes colma nuestro ser
su gracia no ha de fenecer.

tr. Federico J. Pagura

6b | Be present at our table, Lord (Table graces)

English

Be present at our table, Lord;
be here and everywhere adored.
This food now bless, and grant that we
may strengthened for your service be.

John Cennick (altd.)

German

1 Herr, gib uns unser täglich Brot.
Lass uns bereit sein, in der Not
zu teilen, was du uns gewährt.
Dein ist die Erde, die uns nährt.

2 Herr, du bist unser täglich Brot.
Du teilst dich aus in deinem Tod.
Wir loben dich und danken dir.
Aus deiner Liebe leben wir.

Edwin Nievergelt

French

A notre table, viens, Seigneur,
Nous t'adorons d'un même cœur.
Que tous les humains soient bénis
pour te fêter en Paradis.

tr. Marc Chambron

Spanish

A nuestra mesa ven, Señor
Te recibimos con amor.
Bendice, ahora, nuestro pan
y de tu gracia haz nos vivir.

tr. Simei Monteiro

Alleluia

Orthodox Liturgy, Kiev, Ukraine

Al - le - lu - i - a, Al - le - lu - i - a,

Al - - - - le - lu - i - a.

8

asato maa sad gamaya

Trad. Sanskrit, India

In slow, free time

1st time LEADER, 2nd time ALL

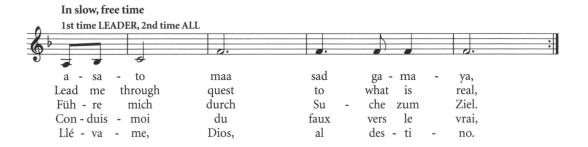

a - sa - to	maa	sad	ga - ma - ya,
Lead me through	quest	to	what is real,
Füh - re mich durch	Su -	che zum	Ziel.
Con - duis - moi du	faux	vers le	vrai,
Llé - va - me, Dios,	al	des - ti -	no.

ta - ma - so	maa_____	jyo - tir	ga - ma - ya,
from the dark	night_____	guide me	to the light.
In dunk-ler	Nacht_____	lass Licht	mich se - hen.
de l'ob-scu - ri - té____	vers la	lu - miè - re,
Guí - a - me,	oh Dios,_	a la	cla - ra luz.

mri - tyor_____	maa____	am - ri - tam_____	ga - ma - ya.
From death_____	take me_	to the realms____	of liv - ing.
Vom Tod_____	führ mich	in das Land____	des Le - bens.
et de no - tre	mort____	vers la vie____	en - tiè - re.
De la muer - te	con-dú - ce-me	a_____	la vi - da.

English tr. Fred Kaan; German tr. Dietrich Werner; French tr. Marc Chambron; Spanish tr. Martin Junge

9

Bendice, Señor, nuestro pan

Unknown, Argentina

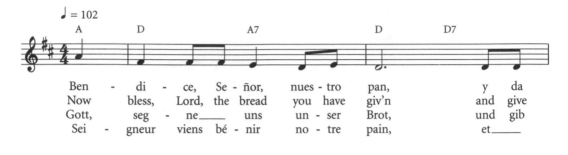

Ben - di - ce, Se - ñor, nues - tro pan, y da
Now bless, Lord, the bread you have giv'n and give
Gott, seg - ne___ uns un - ser Brot, und gib
Sei - gneur viens bé - nir no - tre pain, et___

pan a los que tien - en___ ham - bre___ y ham-bre de jus - ti-cia a los que tien - en
bread to this our world that is hun-gry; to those with bread give hun-ger for true jus-tice
al - len Brot, die hung - rig___ sind, und gib Hun - ger nach Ge - rech-tig-keit den Sat -
nour - ris tous ceux qui ont faim. Don-ne faim___ de jus - tice à ceux qui ont du

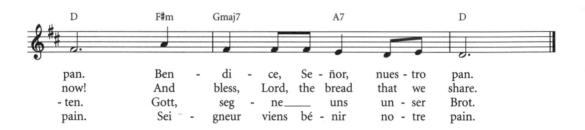

pan. Ben - di - ce, Se - ñor, nues - tro pan.
now! And bless, Lord, the bread that we share.
- ten. Gott, seg - ne___ uns un - ser Brot.
pain. Sei - gneur viens bé - nir no - tre pain.

English tr. Fred Kaan; German tr. Dieter Trautwein; French tr. Marc Chambron

Bwana awabariki

Trad. Swahili, Tanzania

$\quad$ = 86

Bwa - na___ a - wa - ba - ri - ki, Bwa - na___ a - wa - ba - ri - ki,
May___ God___ grant you a bless-ing, may___ God___ grant you a bless-ing,
Got - tes___ Se - gen sei mit dir, Got - tes___ Se - gen sei mit dir,
Que___ Dieu___ vous bé - nis-se tous, Que no - tre Dieu vous bé - nis-se tous.

Bwa - na___ a - wa - ba - ri - ki mi - le - le.
may___ God___ grant you a bless - ing ev - er - more.
Got - tes___ Se - gen sei mit dir je - den Tag.
Que___ Dieu___ vous bé - nis-se tous, et tou - jours.

LEADER

U - ki - mcha Bwa - na.
Re - vere the Lord.___
Eh - re sei Gott.___
Lou - ez le Sei - gneur.

ALL

Bwa - na___ a - wa - ba - ri - ki.
May___ God___ grant you a bless - ing.
Got - tes___ Se - gen sei mit dir.
Et___ que Dieu vous bé - nis-se tous.

English tr. unknown; German tr. Dietrich Werner; French tr. Marc Chambron

11 | Bí, a Íosa, im chroíse

Anon.

Trad., Ireland
arr. Anthony F. Carver

Gaelic

1 Bí, a Íosa, im chroíse i gcuimhne gach uair,
   bí, a Íosa, im chroíse le haithrí go luath,
   bí, a Íosa, im chroíse le cumann go buan,
   ó, a Íosa, 'Dhé dhílis, ná scar thusa uaim.

2 'Sé, a Íosa mo ríse, mo chara is mo ghrá;
   'sé, a Íosa mo dhídean ar pheacaí is ar bhás;
   'sé, a Íosa mo aoibhneas, mo scáthán de ghnáth;
   is a Íosa, 'Dhé dhílis, ná scar uaim go brách.

3 Bí, a Íosa, go síoraí im chroí is im bhéal,
   bí, a Íosa, go síoraí im thuigse mar an gcéann',
   bí, a Íosa, go síoraí im mheabhair mar léann,
   's ó, a Íosa, 'Dhé dhílis, ná fág mé liom féin.

English

1 O Jesus, every moment be in my heart and mind;
   O Jesus, stir my heart with sorrow for my sins;
   O Jesus, fill my heart with never-failing love;
   O Jesus, sweet Master, never part from me.

2 O Jesus is my king, my friend, and my love;
   O Jesus is my shelter from sinning and from death;
   O Jesus, my gladness, my mirror all my days;
   O Jesus, my dearest Lord, never part from me.

3 O Jesus, be for ever in my heart and my mouth;
   O Jesus, be for ever in my thought and my prayer;
   O Jesus, be for ever in my quiet mind;
   O Jesus, my sweet Lord, do not leave me alone.
                 tr. George Otto Simms

German

1 O Jesus, alle Zeiten sei du in meinem Herz.
   O Jesus, mit Reue lass mich sehen meine Sünd'.
   O Jesus, meinen Frieden in deiner Liebe find'.
   O Jesus, Herr und Bruder, an meiner Seite bleib fest.

2 O Jesus, mein König, mein Freund und mein Halt.
   O Jesus, der Schutz gibt vor Abgrund und Sünd'.
   O Jesus, meine Freude, meine Hoffnung, jeden Tag.
   O Jesus, Herr und Bruder, an meiner Seite bleib fest.

3 O Jesus, meine Worte, meine Sinne lenke du.
   O Jesus, meiner Seele, aufgewühlt, schenke Ruh.
   O Jesus, meinem Denken, meinem Beten, gebe Ziel.
   O Jesus, Herr und Bruder, an meiner Seite bleib fest.
                 tr. Dietrich Werner

12 Cantai ao Senhor

based on Psalm 98

Unknown, Brazil
arr. Simei Monteiro

No percussion is used in this song.

(*verses overleaf*)

Portuguese

1 Cantai ao Senhor um cântico novo,
  cantai ao Senhor um cântico novo,
  cantai ao Senhor um cântico novo,
  cantai ao Senhor, cantai ao Senhor!

2 Porque ele fez, e faz maravilhas, (x3)
  cantai ao Senhor, cantai ao Senhor!

3 Cantai ao Senhor, bendizei o seu nome. (x3)
  Cantai ao Senhor, cantai ao Senhor!

4 É ele quem dá o Espírito Santo. (x3)
  Cantai ao Senhor, cantai ao Senhor!

5 Jesus é o Senhor! Amén, aleluia! (x3)
  Cantai ao Senhor, cantai ao Senhor!

Spanish

1 Cantad al Señor un cántico nuevo,
  cantad al Señor un cántico nuevo,
  cantad al Señor un cántico nuevo,
  cantad al Señor, cantad al Señor.

2 Porque el creó, creó maravillas: (x3)
  cantad al Señor, cantad al Señor.

3 La peña se abrió, y agua bebimos: (x3)
  cantad al Señor, cantad al Señor.

4 El es quien nos da el Espíritu Santo: (x3)
  cantad al Señor, cantad al Señor.

5 ¡Jesus es Señor! ¡Amén, aleluya! (x3)
  Cantad al Señor, cantad al Señor.

English

1 O sing to the Lord, O sing God a new song.
  O sing to the Lord, O sing God a new song.
  O sing to the Lord, O sing God a new song.
  O sing to our God, O sing to our God.

2 By this holy power our God has done wonders. (x3)
  O sing to our God, O sing to our God.

3 So dance for our God and blow all the trumpets. (x3)
  O sing to our God, O sing to our God.

4 O shout to our God, who gave us the Spirit. (x3)
  O sing to our God, O sing to our God.

5 For Jesus is Lord! Amen, alleluia. (x3)
  O sing to our God, O sing to our God.
                                tr. Gerhard Cartford

German

1 Singt Gott, unserm Herrn, singt ihm neue Lieder!
  Singt Gott, unserm Herrn, singt ihm neue Lieder!
  Singt Gott, unserm Herrn, singt ihm neue Lieder!
  Singt Gott, unserm Herrn, singt Gott, unserm Herrn!

2 Denn Wunder tat Gott, er tut sie noch immer. (x3)
  Singt Gott, unserm Herrn, singt Gott, unserm Herrn!

3 Jauchzt Gott, alle Welt, lobsingt seinem Namen! (x3)
  Singt Gott, unserm Herrn, singt Gott, unserm Herrn!

4 Gott gib uns den Geist, der aufweckt und sammelt! (x3)
  Singt Gott, unserm Herrn, singt Gott, unserm Herrn!

5 Denn Jesus ist der Herr. Amen, halleluja! (x3)
  Singt Gott, unserm Herrn, singt Gott, unserm Herrn!
  tr. Dorival Ristoff and Dieter Trautwein

French

1 Chantons au Seigneur un nouveau cantique.
  Chantons au Seigneur un nouveau cantique.
  Chantons au Seigneur un nouveau cantique.
  Chantons au Seigneur, chantons au Seigneur.

2 C'est lui, et lui seul qui fait des merveilles. (3x)
  Chantons au Seigneur, chantons au Seigneur.

3 Bénissez le nom du Seigneur du monde. (3x)
  Chantons au Seigneur, chantons au Seigneur.

4 Que le Saint-Esprit conduise la danse. (3x)
  Chantons au Seigneur, chantons au Seigneur.

5 Jésus est Seigneur, amen, alléluia. (3x)
  Chantons au Seigneur, chantons au Seigneur.
  tr. Marc Chambron

13 | Come, Holy Spirit, descend on us

Words and music by
the Wild Goose Worship Group, Scotland

English

1 Come, Holy Spirit, descend on us,
    descend on us.
  We gather here in Jesus' name.

2 Come, Breath of Heaven, …

3 Come, Word of Mercy, …

4 Come, Fire of Judgement, …

5 Come, Great Creator, …

6 Come to unite us, …

7 Come to disturb us, …

8 Come to inspire us, …

German

1 Komm, Geist des Lebens, komm über uns,
    komm über uns,
  und unter uns wächst Gottes Reich.

2 Komm, Brot des Himmels, …

3 Komm, du Erbarmen, …

4 Komm, Flamme Gottes, …

5 Komm, grosser Schöpfer, …

6 Komm, zu versöhnen, …

7 Komm, störe uns auf, …

8 Komm, gehe uns auf, …
    tr. Thomas Laubach

French

1 Viens, Esprit Saint, viens, descend sur nous,
    descend sur nous.
  Rassemble nous en Jésus-Christ.

2 Viens, souffle divin, …

3 Viens, verbe de grâce, …

4 Viens, feu du jugement, …

5 Viens, puissant créateur, …

6 Viens pour nous unir, …

7 Viens nous déranger, …

8 Viens nous inspirer, …
    tr. Hans Schmocker

14 Crashing waters at creation

Sylvia Dunstan

CRASHING WATERS 87 87
William A. Cross, Canada

Crash - ing___ wa - ters at cre - a - tion,
To - sen - de Was - ser, im Akt der___ Schöp - fung,

or - dered by the___ Spi - rit's breath, first to___ wit - ness
fin - den___ Ruh durch des Schöp - fers Kraft, Him - mel und Er - de,

day's be - gin - ning from the___ bright - ness of___ night's death.
A - bend und Mor - gen, wur - den___ durch des Schöp - fers Macht.

English

1 Crashing waters at creation,
  ordered by the Spirit's breath,
  first to witness day's beginning
  from the brightness of night's death.

2 Parting water stood and trembled
  as the captives passed on through,
  washing off the chains of bondage
  channel to a life made new.

3 Cleansing water once at Jordan
  closed around the one foretold,
  opened to reveal the glory
  ever new and ever old.

4 Living water, never ending,
  quench the thirst and flood the soul.
  Wellspring, Source of life eternal,
  drench our dryness, make us whole.

German

1 Tosende Wasser, im Akt der Schöpfung,
  finden Ruh durch des Schöpfers Kraft,
  Himmel und Erde, Abend und Morgen,
  wurden durch des Schöpfers Macht.

2 Wasser des Schilfmeers standen zur Seite,
  Israels Durchzug zu bereiten.
  Ketten der Knechtschaft, aufzubrechen,
  Wege zur Freiheit öffnen stets neu.

3 Wasser des Jordans, im Bad der Taufe,
  umschlossen den, der ist Gottes Sohn,
  er, Gottes Zeichen, Träger des Geistes,
  Gottes Lieb' er zeiget schon.

4 Wasser des Lebens, fliesst ohne Ende,
  still uns'ren Durst und erfrisch uns all,
  Quell ew'gen Lebens, Weltenwende,
  erfüll die Schöpfung, mach uns heil.
  tr. Dietrich Werner

15

Chimwemwe mwa Yesu

Trad. Nyanja, Zambia

♩ = 120

LEADER ... ALL

1. Chi - mwe - mwe mwa Ye - su, chi - mwe - mwe mwa Ye - su, chi -
2. U - mo - zi mwa Ye - su, u - mo - zi mwa Ye - su, u -
3. Mu - te - nde mwa Ye - su, mu - te - nde mwa Ye - su, mu -

LEADER ALL

- mwe - mwe mwa Ye - su, chi - mwe - mwe mwa Ye - su, chi - mwe - mwe mwa Ye - su, chi -
- mo - zi mwa Ye - su, u - mo - zi mwa Ye - su, u - mo - zi mwa Ye - su, u -
- te - nde mwa Ye - su, mu - te - nde mwa Ye - su, mu - te - nde mwa Ye - su, mu -

LEADER ALL

- mwe - mwe mwa Ye - su, chi - mwe - mwe mwa Ye - su.
- mo - zi mwa Ye - su, u - mo - zi mwa Ye - su.
- te - nde mwa Ye - su, mu - te - nde mwa Ye - su.

(1. Rejoice in Jesus. 2. We're one in Jesus. 3. Peace in Jesus.)

(1. Freut euch an Jesus. 2. Wir sind einig in Jesus. 3. Friede in Jesus.)

(1. Réjouissez-vous en Jésus. 2. Nous sommes un en Jésus. 3. Paix en Jésus.)

(1. Regocijémonos en Jesús. 2. Somos uno en Jesús. 3. Paz en Jesús.)

16 Du är helig

Words and music by
Per Harling, Sweden

English tr. Per Harling; German tr. Wolfgang Leyk (altd.); French tr. Joëlle Gouël; Spanish tr. Rachel Achon

1 och 2 kan sjungas samtidigt.

Parts 1 and 2 can be sung at the same time.

Teile 1 und 2 können gleichzeitig gesungen werden.

Les deux parties peuvent se chanter en même temps.

Las partes 1 y 2 se pueden cantar al mismo tiempo.

17 | Du satte dig selv i de nederstes sted

Words and music by
Merete Wendler, Denmark

♩ = 120

Du sat - te dig selv i de ne - der-stes sted._ Du sat - te dig
You came, took your place with the hum-ble and poor._ You came and took
Du hast dich ge - stellt zu den Ar-men am Rande._ Du hast an - ge -

op mod de sto-res for - træd._ Du sat - te dig ind i de y - der-stes nød.
is - sue with sta-tus and power._ You came and de - scend - ed to depths of des-pair.
- fragt die vom ho - hen Stande. Du hast dich ver - letz - lich ge-macht bis zum Tod.

_ Du sat - te dig ud o - ver døds-angst og død._
_ You faced death and dy - ing—its hor - ror and scare._
_ Du hast_ ge - tragen al - ler Men - schen Not._

Danish

1 Du satte dig selv i de nederstes sted.
  Du satte dig op mod de stores fortræd.
  Du satte dig ind i de yderstes nød.
  Du satte dig ud over dødsangst og død.

2 Nu taler du til os fra højeste sted,
  hvor ingen kan udsätte dig for fortræd,
  og lever og lider dog med os endnu,
  så ingen er mere i live end du.

3 Du kæmper i verden for frihed og fred.
  Når andre gir op, blir du viljefast ved.
  Du følger os ind i den yderste nød,
  og kalder os ud af den inderste død.

4 Du sender os ned i de nederstes sted.
  Du sätter os op mod de stores fortræd.
  Du lever i os i de inderste lag.
  Vi lever i dig på den yderste dag.

English

1 You came, took your place with the humble and poor.
  You came and took issue with status and power.
  You came and descended to depths of despair.
  You faced death and dying—its horror and scare.

2 Today you are with us in 'words from great height',
  where nothing can stifle the reach of your light;
  you stay with your people in sickness and health,
  thus none is more truly alive than your Self.

3 You share the world's struggle for freedom and peace,
  and never give up until all are released.
  You follow us into the depths of despair
  and raise us to life from what death would ensnare.

4 You send us to sit with the humble and poor,
  and—with them—rise up against wielders of power.
  You, deep in our being, bring hope to our hearts;
  in you we're alive until earth-light departs.
  tr. Fred Kaan

German

1 Du hast dich gestellt zu den Armen am Rande.
  Du hast angefragt die vom hohen Stande.
  Du hast dich verletzlich gemacht bis zum Tod.
  Du hast getragen aller Menschen Not.

2 Du bist uns nah, wenn wir nah sind den Armen.
  Wo's dunkel ist, kann doch dein Licht nicht ersticken.
  Ob Krankheit drückt, Schmerz oder Sterben sich anschickt,
  lebendig wirkst du als uns're heilende Kraft.

3 Du gibst nicht auf, bis alle Welt findet Frieden.
  Du lässt nicht nach, bis wer gefangen ist frei.
  Du lässt uns in tiefster Not nicht allein,
  willst Retter uns und aller Heiland sein.

4 Du sendest uns zu Orten, wo Menschen leiden,
  den Aufstand zu proben, neues Leben bereiten.
  Tief in uns schaffst Du selbst unsere Hoffnung.
  In dir wir leben bis zu unserem End.
  tr. Dietrich Werner

Dvasia Šventoji, ateiki

Veni Sancte Spiritus
tr. M. Mažvydas

Trad., Lithuania
arr. Rimantas Sliužinskas

Dva - sia Šven - to - ji, a - tei - ki ir
O ho - ly, lov - ing Spi - rit, come, give
Komm Heil' - ger Geist, ent - zün - de uns, lass

iš dan - gaus mums nu - švies - ki, švie -
sight to blind and voice to dumb, give
sehn die Blin - den, öffne den Mund, die

- sy - bės sa - vo spin - du - lį! A -
hope, so we may not suc - cumb. Al -
Hoff - nung lass nicht un - ter - gehn. Hal -

- le - liu - ja! A - le - liu - ja!
- le - lu - ia! Al - le - lu - ia!
- le - lu - ja! Hal - le - lu - ja!

Lithuanian

1 Dvasia Šventoji, ateiki
  ir iš dangaus mums nušvieski,
  šviesybės savo spindulį!
  Aleliuja! Aleliuja!

2 Ateik, ak Tėve vargdienių.
  Ateik, davėja dovanų!
  Ateik, šviesybe širdelių!
  Aleliuja! Aleliuja!

3 Linksmintoja geriausioji,
  viešnia laukta širdžių esi,
  saldų darai mums atvėsį.
  Aleliuja! Aleliuja!

4 Tu darbuose pailsini,
  o troškume atvėsini,
  ir verkiantį palinksmini!
  Aleliuja! Aleliuja!

English

1 O holy, loving Spirit, come,
  give sight to blind and voice to dumb,
  give hope, so we may not succumb.
  Alleluia! Alleluia!

2 Come, God of gifts, empower us,
  come, Light of Light, enlighten us,
  come, God of truth, deliver us.
  Alleluia! Alleluia!

3 O Spirit, fire, wind and dove,
  give light to all from heav'n above,
  give every longing heart your love.
  Alleluia! Alleluia!

4 O Spirit, purify our soul,
  come, healing wind, and make us whole,
  give joy, give faith and be our goal.
  Alleluia! Alleluia!

tr. Per Harling

German

1 Komm Heil'ger Geist, entzünde uns,
  lass sehn die Blinden, öffne den Mund,
  die Hoffnung lass nicht untergehn.
  Halleluja! Halleluja!

2 Komm Heil'ger Geist, belebe uns,
  die Gaben des Geistes, die schenke uns,
  zur Wahrheit du berufe uns.
  Halleluja! Halleluja!

3 Komm Heil'ger Geist, verwandle uns.
  Geist, Feuer und Wind vermehr geschwind,
  mit deiner Lieb rühr jeden an.
  Halleluja! Halleluja!

4 Komm Heil'ger Geist, und mach uns heil.
  Versöhne uns in aller Welt.
  Versöhnte Kirche lass uns sein.
  Halleluja! Halleluja!

tr. Dietrich Werner

19 | Du bist unser alles

St Columbanus

Sr. Adelheid Wenzelmann, Germany

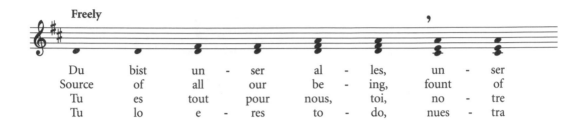

Du bist un - ser al - les, un - ser
Source of all our be - ing, fount of
Tu es tout pour nous, toi, no - tre
Tu lo e - res to - do, nues - tra

Le - ben, un - ser Licht, un - ser Heil,
liv - ing, you are light, heal - ing power!
vie, et no - tre jour, no - tre pain
vi - da, nues - tra luz, la sa - lud,

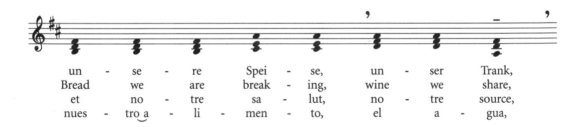

un - se - re Spei - se, un - ser Trank,
Bread we are break - ing, wine we share,
et no - tre sa - lut, no - tre source,
nues - tro a - li - men - to, el a - gua,

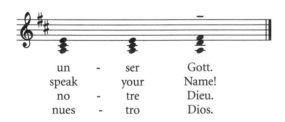

un - ser Gott.
speak your Name!
no - tre Dieu.
nues - tro Dios.

English tr. Fred Kaan; French tr. Marc Chambron; Spanish tr. Martin Junge

20 Eat this bread

based on John 6: 47–58

Jacques Berthier
for the Taizé Community, France

Eat this bread, drink this cup, come to me and nev - er be hun - gry.
Prends ce pain, bois ce vin, viens à moi, et tu n'au - ras plus faim.
Esst das Brot, trinkt den Wein, kommt zu mir, seid nim-mer-mehr hun - grig!

Eat this bread, drink this cup, trust in me and you will not thirst.
Manges et bois, crois en moi, et ma paix se - ra a - vec toi.
Esst das Brot, trinkt den Wein, glaubt an mich, seid nim-mer-mehr durstig!

(Come este pan, bebe esta copa, ven a mi y no tengas nunca hambre. Come este pan, bebe esta copa, confía en mi y no pasarás sed.)

May be sung as ostinato, or alternating with the verses on the next page.

21

E toru nga mea

based on 1 Corinthians 13: 13

Maori, Aotearoa New Zealand

♩. = 50

LEADER

E   to - ru nga   mea_____   Nga mea   nu
Of   all things that are_____   three mat - ter
Nun a -   ber   bleiben_____   drei teu -   re

ALL

E   to - ru   nga   mea
Of   all things that   are
Nun a -   ber bleiben

nui_____   E kii   a - na_____   Te Pai - pe -
most._____   They are   des - cribed_____   in Ho - ly
Worte_____   der Bi - bel,   Worte_____   für uns   ge -

Nga mea   nu nui   E kii   a - na
three mat - ter most.   They are   des-cribed
drei teu - re Worte   der Bi - bel, Worte

- ra_____   Wha - ka - po - no_____   Tu-ma - na -
Writ:_____   the   first   of them: faith!_____   the se - cond:
- geben:_____   Glaub', Hoff-nung und   Liebe,_____   a - ber die

Te Pai - pe - ra   Wha - ka - po - no
in Ho - ly Writ: the   first   of them: faith!
für uns   ge-geben: Glaub', Hoff-nung und Liebe,

English tr. Fred Kaan; German tr. Dietrich Werner

(Il y a trois choses essentielles. C'est ce que disent les mots vivants de la Bible: la foi, l'espérance, mais la plus grande, c'est l'amour.)

(Tres palabras son valiosas, la Biblia las mantiene vivas. Una es la fe, la otra esperanza. Pero la mayor de ellas es amor.)

22 El cielo canta alegría

Words and music by
Pablo Sosa, Argentina

El cie-lo can-ta a-le-grí-a, ¡A - le - lu - ya!
Hea-ven is sing-ing for joy,___ Al - le - lu - ia!
Him-mel er-klin-gen vor Freu-de, Hal-le - lu - ja!

por-que en tu vi-da y la mí-a bri-lla la glo - ria de Dios.
for in your life and in mine is shin-ing the glo - ry of God.
Denn dein und mein Le-ben glän-zen im_ Licht, das Gott rühmt und ehrt.

¡A - le - lu - ya! ¡A - le - lu - ya!
Al - le - lu - ia! Al - le - lu - ia!
Hal - le - lu - ja! Hal - le - lu - ja!

¡A - le - lu - ya! ¡A - le - lu - ya!
Al - le - lu - ia! Al - le - lu - ia!
Hal - le - lu - ja! Hal - le - lu - ja!

Carnavalito

SHAKERS

CONGAS

GUITARS

BASS

Instruments: charango and quenas (or guitar and recorder/flute)

Spanish

1 El cielo canta‿alegría, ¡Aleluya!
  porque‿en tu vida‿y la mía brilla la gloria de Dios.
  ¡Aleluya! ¡Aleluya! ¡Aleluya! ¡Aleluya!

2 El cielo canta‿alegría, ¡Aleluya!
  porque‿a tu vida‿y la mía las une‿el amor de Dios.

3 El cielo canta‿alegría, ¡Aleluya!
  porque tu vida‿y la mía proclamarán al Señor.

English

1 Heaven is singing for joy, Alleluia!
  for in your life and in mine is shining the glory of God.
  Alleluia! Alleluia! Alleluia! Alleluia!

2 Heaven is singing for joy, Alleluia!
  for your life and mine unite in the love of the Lord.

3 Heaven is singing for joy, Alleluia!
  for your life and mine will always bear witness to God.
        tr. Pablo Sosa

German

1 Himmel erklingen vor Freude, Halleluja!
  Denn dein und mein Leben glänzen im Licht, das Gott rühmt und ehrt.
  Halleluja! Halleluja! Halleluja! Halleluja!

2 Himmel erklingen vor Freude, Halleluja!
  Denn dein und mein Leben sind durch die Liebe Gottes vereint.

3 Himmel erklingen vor Freude, Halleluja!
  Denn dein und mein Leben können ein Zeugnis geben für Gott.
        tr. Dieter Trautwein

23 Fader vår

Matthew 6: 9–13

Egil Hovland, Norway

Fa-der vår,   du som er i him-mel-en.   La ditt navn hol-des   hel - lig.
Our   Fa - ther in   hea - ven,   hal-lowed be   your_   name, your
Va - ter   un - ser im   Him - mel.   Hei - lig sei dein_   Na - me.

La ditt ri - ke   kom - me.__   La din vil - je   skje   på   jor-den som i
king - dom_   come, your_   will_   be__   done on   earth_   as in
Dein_   Reich   kom - me. Dein   Wil - le ge - scheh auf   Er - den wie im

him-mel-en. Gi oss i dag vårt dag - li - ge brød. For - lat oss vår skyld, som
hea - ven. Give us to - day our dai - ly___ bread. For - give us our sins as
Him - mel. Gib heu-te un - ser täg - li-ches Brot. Ver - gib uns'-re Schuld, wie

vi___ og for - la - ter vå - re skyld-ne - re. Led oss ik - ke inn i___
we for - give___ those who sin a - gainst us. Lead us not in - to temp -
wir___ ver - ge - ben uns'-ren Schul-di-gern. Füh - re uns nicht in Ver -

fris - tel - se, men frels oss fra det on - de. For___ ri - ket er
-ta - tion but de - li - ver us from e - vil. For the king - dom, the
-su - chung, er - lös uns von dem Bö - sen. Denn das Reich und die

ditt, og___ mak - ten og æ - ren nå og i e - vig - het. A - men.
power and the glo - ry are yours___ now and for e - ver. A - men.
Kraft und die Herr-lich-keit sind dein___ jetzt und in E - wig - keit. A - men.

24 For God alone

Words and music by
Eleanor Daley, Canada

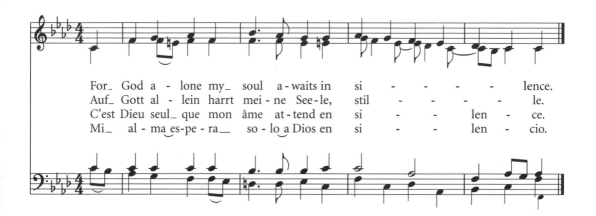

For_ God a - lone my_ soul a - waits in si - - - - - - lence.
Auf_ Gott al - lein harrt mei - ne See - le, stil - - - - le.
C'est Dieu seul_ que mon âme at - tend en si - - len - ce.
Mi_ al - ma es - pe - ra_ so - lo a Dios en si - - len - cio.

German tr. Dorothea Wulfhorst; French tr. Marc Chambron; Spanish tr. Martin Junge

© Harold Flammer Inc./Music Sales

25

For the healing of the nations

Fred Kaan, Netherlands/England

WESTMINSTER ABBEY 87 87 87
Henry Purcell, England

For the heal - ing_ of the na - tions, God, we pray___ with
Für die Hei - lung al - ler Völ - ker_ bit - ten wir___ mit
Veuil - le gué - rir_ tous les peu - ples, O Sei - gneur,_ nous

one_ ac - cord; for a just and_ e - qual shar - ing
ei - nem_ Mund um ge - rech - tes,_ glei - ches Tei - len
t'en pri - ons; ap - prends - nous le_ vrai_ par - ta - ge

of___ the_ things___ that earth af - fords. To a life of
auf___ dem glei - chen Er - den - rund. Hilf, dass wir in
des___ biens de___ ta cré - a - tion. Et que ton a -

love in ac - tion help us rise and pledge_ our_ word.
tät' - ger Lie - be wu - chern mit dem eig - nen_ Pfund.
- mour in - spi - re nos pa - roles et nos_ ac - tions.

English

1 For the healing of the nations,
  God, we pray with one accord;
  for a just and equal sharing
  of the things that earth affords.
  To a life of love in action
  help us rise and pledge our word.

2 Lead us forward into freedom,
  from despair your world release;
  that, redeemed from war and hatred,
  all may come and go in peace.
  Show us how through care and goodness
  fear will die and hope increase.

3 All that kills abundant living,
  let it from the earth be banned:
  pride of status, race or schooling,
  dogmas that obscure your plan.
  In our common quest for justice
  may we hallow life's brief span.

4 You, Creator God, have written
  your great name on humankind;
  for our growing in your likeness
  bring the life of Christ to mind;
  that by our response and service
  earth its destiny may find.

German

1 Für die Heilung aller Völker
  bitten wir mit einem Mund
  um gerechtes, gleiches Teilen
  auf dem gleichen Erdenrund.
  Hilf, dass wir in tät'ger Liebe
  wuchern mit dem eignen Pfund.

2 Führ uns, Vater, in die Freiheit,
  mach uns von Verzweiflung frei,
  dass erlöst von Hass und Kriegen
  Friede mit uns allen sei.
  Zeig uns, wie durch Hilf und Güte
  Angst stirbt, Hoffnung wächst herbei.

3 Alles, was das Leben tötet,
  stelle unter deinen Bann:
  Stolz auf Stellung, Farbe, Klasse,
  Lehren gegen deinen Plan.
  Noch im Kampf für das, was recht ist,
  sehn wir Leben heilig an.

4 Schöpfer, du schreibst deinen Namen
  tief ins Buch der Menschheit ein;
  lass in uns dein Bildnis wachsen,
  hilf uns, Christus näher sein,
  dass durch uns'res Lebens Antwort
  Erde glänzt in deinem Schein.
  tr. Dieter Trautwein

French

1 Veuille guérir tous les peuples,
  O Seigneur, nous t'en prions;
  apprends-nous le vrai partage
  des biens de ta création.
  Et que ton amour inspire
  nos paroles et nos actions.

2 Conduis-nous, conduis le monde,
  vers la joie, la liberté;
  sans les haines, sans les guerres,
  que tous puissent vivre en paix.
  Aide-nous à mieux répandre
  l'espérance et la bonté.

3 Que la vie reste première,
  que soit banni ce qui tue;
  que l'orgueil, la suffisance,
  l'argent ne dominent plus.
  Que la justice progresse,
  qu'elle soit partout vécue.

4 Ton nom, Seigneur, d'âge en âge,
  est inscrit sur les humains;
  tous, créés à ton image,
  c'est le Christ qui nous fait tiens.
  Que, par notre obéissance,
  le monde aille à son destin.
  tr. Marc Chambron

26

Gloria a Dios

Unknown, Peru

Andean instruments: charango, zampoñas, quenas, percussion

Spanish

1 Gloria a Dios, gloria a Dios, gloria en los cielos.
  A Dios la gloria por siempre.
  Aleluya, amen. Aleluya, amen. Aleluya, amen.

2 Gloria a Dios, gloria a Dios, gloria a Jesucristo.

3 Gloria a Dios, gloria a Dios, gloria al Espíritu Santo.

English

1 Glory to God, glory to God, glory in the highest.
  To God be glory forever.
  Alleluia, amen. Alleluia, amen. Alleluia, amen.

2 Glory to God, glory to God, glory to Christ Jesus.

3 Glory to God, glory to God, glory to the Spirit.

German

1 Ehre sei Gott, Ehre sei Gott, Ehre in der Höhe.
  Gebt Gott die Ehre für immer.
  Halleluja, amen. Halleluja, amen. Halleluja, amen.

2 Ehre sei Gott, Ehre sei Gott, Ehr' sei Jesus Christus.

3 Ehre sei Gott, Ehre sei Gott, Ehr' dem Heil'gen Geiste.
  tr. Dieter Trautwein

French

1 Gloire à Dieu, gloire à Dieu, à Dieu, gloire dans les hauts lieux.
  Gloire au Seigneur, gloire à jamais.
  Alléluia, amen. Alléluia, amen. Alléluia, amen.

2 Gloire à Dieu, gloire à Dieu, à Dieu, gloire au Seigneur Jésus.

3 Gloire à Dieu, gloire à Dieu, à Dieu, gloire à l'Esprit très saint.
  tr. Joëlle Gouël

27

Gloria in excelsis Deo!

Luke 2: 14

GLORIA III
Jacques Berthier
for the Taizé Community, France

Canon

♩. = 126

INSTRUMENTS (ad lib.)

Dm Gm C F Dm Gm C F
2.

Glo - ri - a, glo - ri - a, in ex - cel - sis De - o!

Dm Gm C F Dm Gm C F
3. 4.

Glo - ri - a, glo - ri - a, al - le - lu - ia, al - le - lu - ia!

(Glory to God in the highest!)

(Ehre sei Gott in der Höhe!)

(Gloire à Dieu dans les lieux très-hauts!)

(¡Gloria a Dios en las alturas!)

28 | Gospodi pomilui

from the Orthodox Liturgy, Ukraine

Gos - po - di po - mi - lui, Gos - po - di po - mi - lui,

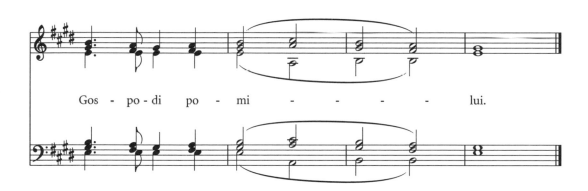

Gos - po - di po - mi - - - - lui.

(Lord, have mercy.)

(Herr, erbarme dich.)

(Seigneur, aie pitié de nous.)

(Señor, ten piedad.)

May be sung by three-part women's or three-part men's voices, or by both together, unaccompanied.

29 God, who made the stars of heaven

Ruth Duck, USA

THIAN BENG   85 85 85 75
I-to Loh, Taiwan

♩ = 100

God, who made the stars of hea-ven, God who spread the earth,
Gott, du machst die Him-mels-ster-ne, machst die Er-de weit.
Toi, qui as cré-é la ter-re le mon-de et les cieux.

breath of ev-ery liv-ing be-ing, fount of life and birth,
A-tem al-ler Le-be-we-sen, Quel-le al-ler Zeit,
Sour-ce de tou-te lu-miè-re, qui bril-le en tout lieu.

you have formed a ser-vant peo-ple, lead us by your hand.
du hast dir ein Volk ge-ru-fen, führst mit gu-ter Hand.
Tu nous fais peu-ple de frè-res, gui-dés par ta main.

Light of na-tions, shine in us; brigh-ten ev - ery_ land.
Licht der Völ - ker, schein in uns ü - ber je - dem Land.
Ser - vi-teurs d'un mê-me Pè-re, peu - ple de té - moins.

English

1 God, who made the stars of heaven,
God who spread the earth,
breath of every living being,
fount of life and birth,
you have formed a servant people,
led us by your hand.
Light of nations, shine in us;
brighten every land.

2 Living Christ, the light of nations,
radiant as the sun,
build us up, a growing body;
knit your church as one.
May our loving be a witness
all the world may see.
Send your Spirit, bond of peace,
source of unity.

3 Spirit God, equip your people,
all with gifts to share:
messengers to speak the gospel,
ministers of care.
So may valleys rise to greatness,
mountains be a plain.
Come, surprise us; change our lives;
heal each heart in pain.

4 So may nations praise your greatness,
do your will on earth,
free the captives from their prisons,
treat the poor with worth.
So may desert, coast and village
sing new songs to you.
Light of nations, fill the world,
making all things new.

(*German and French verses overleaf*)

German

1 Gott, du machst die Himmelssterne,
   machst die Erde weit.
   Atem aller Lebewesen,
   Quelle aller Zeit,
   du hast dir ein Volk gerufen,
   führst mit guter Hand.
   Licht der Völker, schein in uns
   über jedem Land.

2 Christus, Leben, Licht der Völker,
   bist wie Sonnenschein.
   Bau uns auf, ein Volk im Wachsen,
   lass uns einig sein.
   Uns're Liebe sei ein Zeugnis,
   das die Welt kann sehn.
   Schick den Geist als Friedensband,
   aus dem Einheit kommt.

3 Gottesgeist, hilf deinen Leuten,
   teil die Gaben aus;
   wir sind Boten deiner Nachricht,
   dienen deinem Haus.
   Mögen Täler sich erhöhen,
   Berge eben sein.
   Komm, erfüll uns, ändre Leben,
   heile Seelenleid.

4 Mögen Völker dich erhöhen,
   hör'n auf dein Gebot,
   aus den Kerkern Gefang'ne holen,
   lindern Armer Not.
   Mögen Wüste, Küste, Städte
   singen ein neues Lied.
   Licht der Völker, füll die Erde,
   und mach alles neu.

tr. Wolfgang Leyk

French

1 Toi, qui as créé la terre
   le monde et les cieux.
   Source de toute lumière,
   qui brille en tout lieu.
   Tu nous fais peuple de frères,
   guidés par ta main.
   Serviteurs d'un même Père,
   peuple de témoins.

2 Jésus-Christ tu es la vie
   et tu es la joie.
   Tous les croyants tu les lies
   dans la même foi.
   Ton amour qui nous fait vivre,
   le monde le voit.
   Ton Esprit qui nous délivre,
   nous unit en Toi.

3 Esprit-Saint rends-nous dociles
   à tes volontés.
   Pour prêcher ton Evangile
   dans la verité.
   Viens aplanir tous les sentiers,
   hausser les vallées.
   Viens nous aimer et consoler
   les cœurs oppressés.

4 Tous les peuples tu les sauves
   dans le monde entier.
   Tu rends libres tous les pauvres
   et les opprimés.
   C'est ta croix qui est le gage
   d'un amour si grand.
   Ta joie qui nous rend sages
   pour toujours vivants.

tr. Jacques Fischer

30

Haleluya Puji Tuhan

Trad. Soahuku, Indonesia
adapt. Christian I. Tamaela

(Alleluia, praise the Lord.)

(Halleluja, lobet den Herrn.)

(Alléluia, louez le Seigneur.)

(¡Aleluya, alabado sea Dios!)

31

Healing is flowing, deep in the waters

Norman C. Habel, Australia

BUNESSAN 55 54 D
Trad., Scotland

Heal - ing is flow - ing, deep in the wa - ters, heal - ing from
Hei - lung strömt zu uns, durch al - le Was - ser, Hei - lung vom

Gué - ris ce mon - de, Dieu des fon - tai - nes, Dieu de la

E - den, flow-ing from old; all through cre - a - tion, God sends forth
An - fang, Hei-lung für uns. Kräf - te der Hei - lung wir - ken für
sour - ce, des o - cé - ans; net - toie nos fau - tes, la - ve nos

wa - ters, o - ceans of heal - ing, for all the world.
im - mer, in sei - ner gan - zen Schöp-fung für uns.
pei - nes; gué - ris ce mon - de, sois bien-fai - sant.

English

1 Healing is flowing, deep in the waters,
healing from Eden, flowing from old;
all through creation, God sends forth waters,
oceans of healing, for all the world.

2 Healing is rising, fresh with the morning,
healing is rising, bursting with grace;
Christ, as you surge from deep in creation,
heal Earth's deep wounds and rise in this place.

3 Healing is breathing, life from the Spirit,
healing is breathing, making flesh whole;
breath of the Spirit, blow through our churches,
come, heal our people, body and soul.

4 Healing is offered, leaves from the life tree,
healing is offered nations at war;
come, wounded Healer, torn by the violence,
break down the walls, bring peace to our shore.

5 Healing is flowing, free with Christ's body,
healing is flowing, free with Christ's blood;
may this deep power pulse through our bodies,
heal the world's wounds, still bleeding and red.

German

1 Heilung strömt zu uns, durch alle Wasser,
Heilung vom Anfang, Heilung für uns.
Kräfte der Heilung wirken für immer,
in seiner ganzen Schöpfung für uns.

2 Heilung beginnt neu, an jedem Morgen,
wenn frische Kräfte sammeln sich neu.
Tief in der Schöpfung, da wirkst du, Christus,
verbinde Wunden, die Welt mach heil.

3 Heilung heisst atmen im Geist des Lebens,
Körper und Seele finden zur Ruh,
Sende den Geist, Gott, in deine Kirche,
dass sie nicht fürchte Nachteil und Macht.

4 Heilung vollzieht sich nicht ohne Opfer,
Umkehr und Wandel verlangen viel.
Länder im Kriege, lasst euch bekehren,
seht zu dem einen Friedensfürst hin!

5 Heilung fliesst durch uns, die wir verbunden,
in Christi Leib die Kirche wir sind.
Seine Kraft wirke durch unser Leben,
dass wir ein Mittel der Heilung sind.
tr. Dietrich Werner

French

1 Guéris ce monde, Dieu des fontaines,
Dieu de la source, des océans;
nettoie nos fautes, lave nos peines;
guéris ce monde, sois bienfaisant.

2 Guéris ce monde, Dieu qui relève,
et que ta force nous soit donnée;
ta grâce abonde, le Christ se lève;
guéris ce monde par ta bonté.

3 Guéris ce monde, Dieu de la vie,
envoie ton souffle sur toute chair;
qu'il se répande par nos Églises;
guéris ce monde de ses misères.

4 Guéris ce monde, Dieu qui demande,
au lieu des guerres, enfin la paix;
viens mettre un terme à la violence;
guéris ce monde à tout jamais.

5 Guéris ce monde, Dieu qui libère,
par le corps du Christ, et par son sang;
que sa présence nous régénère;
guéris ce monde dès maintenant.
tr. Marc Chambron

32 | Ikke ved makt

Sverre Therkelsen

Knut Nystedt, Norway

Ik - ke ved makt— ik - ke ved vold og ved makt vil du rå - de,
Ne - ver by might, ne - ver by vio-lence or might do you rule us,
Nie-mals mit Macht, nie mit Ge-walt und mit Macht du uns füh - re!

nei, ved din nå - de, ja, ved din kjær - li - ge hånd og din nå - de!
but by your mer - cy, yes, by your mer - ci - ful hand and com - pas - sion.
Gott, uns be - rüh - re! Uns mit der lie - ben - den Hand uns be - rüh - re!

Norwegian

1 Ikke ved makt — ikke ved vold og ved makt vil du råde,
   nei, ved din nåde, ja, ved din kjærlige hånd og din nåde!

2 Ikke ved sverd — ikke ved ild og ved sverd du oss bøye!
   Nei, ved ditt øye, ja, ved et faderlig blikk fra ditt øye.

3 Ikke ved frykt — ikke ved redsel og frykt vil du lede,
   nei, ved din glede, ja, ved det sinn som du fyller med glede!

4 Ikke ved dom — ikke forstøt oss til trelldom og trengsel!
   Nei, gi oss lengsel! Styr våre hjerter ved håp og ved lengsel!

English

1 Never by might, never by violence or might do you rule us,
   but by your mercy, yes, by your merciful hand and compassion.

2 Never with sword, never with fire or with sword do you force us,
   but with your vision, yes, with the gaze of your love do you draw us.

3 Never by fear, never by terror or fear do you drive us,
   but by your gladness, yes, by our souls filled with joy do you lead us.

4 Never by threat, never by menace or threat do you sway us.
   Urge us to seek you, filling our hearts with your hope and your longing.
   tr. Maggie Hamilton

German

1 Niemals mit Macht, nie mit Gewalt und mit Macht du uns führe!
   Gott, uns berühre! Uns mit der liebenden Hand uns berühre!

2 Niemals mit Schwert, niemals mit Flammen und Schwertern uns leite!
   Du uns begleite, uns mit dem Vaterblick zärtlich begleite!

3 Niemals mit Furcht, nie willst mit Gräuel und Furcht du uns biegen.
   Freude wir kriegen! Freiheit und Freude im Herzen wir kriegen.

4 Niemals mit Zwang, spar uns den Zwang und die ängstlichen Tränen!
   Nur lass uns sehnen! Lass nach Erlösung uns zutraulich sehnen!
   tr. Sverre Therkelsen

33

Iloitse, maa! (Halleluja–laulu)

based on Psalm 100

Words and music by
Olli Kortekangas, Finland

Finnish	English

1 Iloitse, maa!
  Herralle riemuhuuto kohottakaa!
  Halleluja, halleluja, halleluja!
  Halleluja, halleluja, halleluja! Aamen.
	1 All the earth, praise!
	  Serve God with heart and soul
	    and sing out your joy!
	  Alleluia, alleluia, alleluia!
	  Alleluia, alleluia, alleluia! Amen.

2 Ylistä, maa!
  Porteilleen kiitoslauluin vaeltakaa!
	2 All the earth's people,
	  enter the gates of God with thanks
	    and with praise!

3 Taivas ja maa
  alati Herran voimaa julistakaa!
	3 Heaven and earth,
	  sing of the power of God, Creator of all!
	  tr. Maggie Hamilton

German	French

1 Jauchze, o Erde!
  Dienet dem Herrn mit Freuden,
    frohlocket ihm!
  Halleluja, halleluja, halleluja!
  Halleluja, halleluja, halleluja! Amen.
	1 Toi, terre entière,
	  acclame le Seigneur, et chante ta joie!
	  Alléluia, alléluia, alléluia!
	  Alléluia, alléluia, alléluia! Amen.

2 Preise, o Erde!
  Kommt durch die Tore jauchzend,
    und danket ihm!
	2 Vous, tous les peuples,
	  venez dans sa maison, pour bénir son nom!

3 Himmel und Erde!
  Preiset den Herrn, den Schöpfer,
    lobt seine Macht!
  tr. Dietrich Werner
	3 Vous, ciel et terre,
	  chantez sa royauté, sa fidélité!
	  tr. Marc Chambron

34 | I will live for you alone

Words and music by
Trisha Watts, Australia

I___ will live for you a-lone, for you a-lone I'll live.
Le - ben will ich dir al-lein, dein ist mein gan - zes Sein.
Pa - ra ti yo vi - vi - ré, tan só - lo pa - ra ti.

Heal me, heal me, heal me and let me live.
Heil mich, heil mich, heil mich und lass mich sein.
Sa - na, sa - na, sa - na mi vi - da ya.

German tr. Dorothea Wulfhorst; Spanish tr. Martin Junge

(Je vivrai pour toi seul. Guéris-moi et fais-moi vivre.)

35

Jeder Teil dieser Erde

Chief Seattle, native American

Stefan Vesper, Germany

Canon

Je - der___ Teil die - ser___ Er - de
Ev - 'ry___ part of this___ earth___
Cha - que___ lieu sur cet - te ter - re
Ca - da lu - gar de es - te mun - do

ist mei - nem Volk___ hei - - - lig.
is to my peo - ple ho - - - ly.
est pour mon peu - ple bien___ sa - cré.
es lu - gar san - to pa - ra mi pueblo.

Je - der___ Teil die - ser___ Er - de
Ev - 'ry___ part of this___ earth___
Cha - que___ lieu sur cet - te ter - re
Ca - da lu - gar de es - te mun - do

ist mei - nem Volk___ hei - - lig.
is to my peo - ple ho - - ly.
est pour mon peu - ple bien sa - cré.
es lu - gar san - to pa - ra mi pueblo.

36

Jesu neem un nu gu shin ga

Seong-Won Park

Sung-Mo Moon, Korea

Je - su neem___ un nu gu shin ga
Je - sus asks, 'Who do you say that I am?'
Du fragst uns Je - sus: Wer bin ich für euch?
Qui di - tes - vous que je suis, moi Jé - sus?

ug ab uee___ khol ja ki uee cha yoo uee kot
Blos - som of___ free - dom in re - pres - sion's dark vale,
Blü - te der___ Frei - heit, mit - ten in Not und Tod,
Fleur de li - ber - té, quand vous___ êtes op - pri - més,

ka nan han beck sung uee chin ku shi ra.
lov - er of jus - tice, and friend of the poor.
Freund al - ler Ar - men und Hei - land der Welt.
je suis l'a - mi des pau - vres et leur sou - tien.

Korean

1 Jesu neem un nu gu shin ga
ug ab uee khol ja ki uee cha yoo uee kot
ka nan han beck sung uee chin ku shi ra.

2 Jesu neem un nu gu shin ga
chul mang uee ku nul so geh
hee mang uee bit
kee ree run beck sung uee mok cha shi ra.

3 Jesu neem un nu gu shin ga
chun chaeng uee yok sa so geh pyung wha uee wang
sum at kin beck sung uee saeng myung iee ra.

English

1 Jesus asks, 'Who do you say that I am?'
Blossom of freedom in repression's
dark vale,
lover of justice, and friend of the poor.

2 Jesus asks, 'Who do you say that I am?'
Beam of hope shining in the night of despair,
guiding and tending all those who are lost.

3 Jesus asks, 'Who do you say that I am?'
Maker of peace in the destruction of war;
to all who suffer the Giver of Life.
tr. Maggie Hamilton

German

1 Du fragst uns Jesus: Wer bin ich für euch?
Blüte der Freiheit, mitten in Not und Tod,
Freund aller Armen und Heiland der Welt.

2 Du fragst uns Jesus: Wer bin ich für euch?
Lichtblick der Hoffnung, trotz Trübsal
und Furcht,
Halt der Verzagten und Hirt der Verlor'n.

3 Du fragst uns Jesus: Wer bin ich für euch?
Du stiftest Frieden überwindest Gewalt.
Du stiftest Leben für alle im Leid.
tr. Dietrich Werner

French

1 Qui dites-vous que je suis, moi Jésus?
Fleur de liberté, quand vous êtes opprimés,
je suis l'ami des pauvres et leur soutien.

2 Qui dites-vous que je suis, moi Jésus?
Lumière d'espoir, éclairant vos ténèbres,
je suis le bon berger, je vous connais.

3 Qui dites-vous que je suis, moi Jésus?
Prince de la paix, malgré toutes vos guerres,
et pour tous les meurtris, je suis la vie.
tr. Marc Chambron

37

Jesus, Lamb of God (Agnus Dei)

Music and additional words by
Marty Haugen, USA

English

1 Jesus, Lamb of God,
   you take away the sins of the world:
   have mercy on us, O Lord.

2 Jesus, Word of God, …
3 Jesus, Bread of Life, …
4 Jesus, Cup of Life, …
5 Jesus, Light of Peace, …
6 Jesus, Hope of all, …
7 Jesus, Word made Flesh, …
8 Jesus, Morning Star, …
9 Jesus, Shepherd King, …
10 Jesus, First and Last, …

11 Jesus, Fire of Love, …
12 Jesus, Risen Lord, …
13 Jesus, Prince of Peace, …
14 Jesus, Breath of Life, …
15 Jesus, Light of Eyes, …
16 Jesus, Rising Son, …
17 Jesus, Bread from Heav'n, …
18 Jesus, Child of Peace, …
19 Jesus, Hope of All, …

German

1 Jesus, Gottes Lamm,
   du nimmst hinweg die Sünde der Welt:
   erbarme dich unser Herr.

2 Jesus, Gottes Wort, …
3 Jesus, Lebensbrot, …
4 Jesus, Lebenstrank, …
5 Jesus, Friedenslicht, …
6 Jesus, Hoffnungsschein, …
7 Fleischgeword'nes Wort, …
8 Jesus, Morgenstern, …
9 Jesus, guter Hirt, …
10 Jesus, A und O, …

11 Jesus, liebevoll, …
12 Auferstand'ner Herr, …
13 Jesus, Friedefürst, …
14 Jesus, Lebenshauch, …
15 Jesus, unser Licht, …
16 Jesus, auferweckt, …
17 Jesus, Himmelsbrot, …
18 Jesus, Friedenskraft, …
19 Hoffnung, stark wie Tod, …

tr. Angelika Joachim

(1. Jésus, Agneau de Dieu, *qui enlèves les péchés du monde: prends pitié de nous.* 2. Jésus, Parole de Dieu, … 3. Jésus, Pain de vie, … 4. Jésus, Coupe de la vie, … 5. Jésus, Lumière de la Paix, … 6. Jésus, Espoir de tous, … 7. Jésus, Parole faite chair, … 8. Jésus, Etoile du matin, … 9. Jésus, Roi des bergers, … 10. Jésus, le Premier et le Dernier, … 11. Jésus, Feu d'amour, … 12. Jésus, Seigneur ressuscité, … 13. Jésus, Prince de la Paix, … 14. Jésus, Souffle de vie, … 15. Jésus, Lumière de nos yeux, … 16. Jésus, Fils ressuscité, … 17. Jésus, Pain du ciel, … 18. Jésus, Enfant de Paix, … 19. Jésus, Espoir de tous, …)

38

Jesus, Lamb of God, bearer of our sin

Words and music by
Bernadette Farrell, England

English

1   Jesus, Lamb of God, bearer of our sin; Jesus, Saviour:
Hear our prayer, hear our prayer; through this bread and wine we share
may we be your sign of peace everywhere.

2   Jesus, Lamb of God, bearer of our pain; Jesus, healer:
3   Jesus, Lamb of God, broken as our bread, here among us:
4   Jesus, Lamb of God, poured out as our wine, shared in gladness:
5   Jesus, Word of God, dwelling with the poor; Jesus, prophet:
6   Jesus, Word of God, dwelling in our midst, Jesus with us:
7   Jesus, Word of God, speaking in our hearts God's compassion:
8   Jesus, Word made flesh, touching each one's need; Jesus, lover:
9   Kneeling by your friends, washing each one's feet; Jesus, servant:
10 Hope beyond despair, dawn of fragile light; Jesus, risen:
11 Tomb of secret hope, open to the dawn; Jesus, living:

German

1   Jesus, Gottes Lamm, du trägst uns're Schuld, Jesus Retter:
Nähre uns, nähre uns, Brot und Wein des neuen Bunds
seien Kraft und Stärkung für das Zeugnis uns.

2   Jesus, Gottes Lamm, du trägst unser Leid, Jesus Heiland:
3   Jesus, Gottes Lamm, du bist unser Brot, wirst gebrochen:
4   Jesus, Gottes Lamm, du bist unser Wein, wirst vergossen:
5   Jesus, Gottes Wort, bist den Armen nah, weist die Wege:
6   Jesus, Gottes Wort, bist uns allen nah, bist Gott-mit-uns:
7   Jesus, Gottes Wort, sprichst in unser Herz Gottes Liebe:
8   Fleischgeword'nes Wort, du kennst uns're Not, bist voll Liebe:
9   Kniest demütig hin, wäscht die Füsse uns, aller Diener:
10 Hoffnung stark wie Tod, erstes Morgenlicht, Auferstand'ner:
11 Grab im Morgenlicht, der, der tot war, lebt, ist erstanden:
        tr. Angelika Joachim

French

1   Christ, Agneau de Dieu, portant nos péchés, tu nous sauves:
Par ce pain, par ce vin, fais de nous, Seigneur Jésus,
des artisans de ta paix, à jamais.

2   Christ, Agneau de Dieu, toi qui viens guérir nos souffrances:
3   Christ, Agneau de Dieu, toi le pain rompu pour les hommes:
4   Christ, Agneau de Dieu, toi le vin versé, joie du monde:
5   Christ, Verbe de Dieu, toi l'ami des pauvres sur terre:
6   Christ, Verbe de Dieu, au milieu de nous, le prophète:
7   Christ, Verbe de Dieu, qui parle a nos cœurs de ton Père:
8   Parole incarnée, toi qui es amour et tendresse:
9   Toi le serviteur, qui viens à genoux, laver nos pieds:
10 Jésus, notre espoir, Christ ressuscité, vraie lumière:
11 Jésus, notre paix, Christ toujours vivant, notre Sauveur:
        tr. Marc Chambron

39

Jézus a jó pásztor

Dóka Zoltán   Trajtler Gábor, Hungary

Jé-zus a jó pász-tor, ve-le me-gyünk. Ö-rök ir-gal-má-ról
Je-sus, Good Shep-herd, guide us through the day. Lead us with lov-ing grace,
Je-sus, der gu-te Hirt, mit dem wir gehen, sei-ne Barm-her-zig-keit,

é - ne-ke-lünk. Nyá-ja-dat ol - tal - mazd,
show us the way. Shel - ter your flock, O Lord,
sie lässt uns singen. Nimm dei-ne Herd' in acht,

nagy a ve-szély! Csen-des nyu-gal-mat adj, ha jön az éj!
your peo-ple hear. Grant us your peace-ful rest when night is near.
wir sind be-droht, gib uns die stil-le Ruh, wenn Nacht sich naht.

Hungarian

1 Jézus a jó pásztor, vele megyünk.
Örök irgalmáról énekelünk.
Nyájadat oltalmazd, nagy a veszély!
Csendes nyugalmat adj, ha jön az éj!

2 Jézus a szőlőtő, ereje nagy.
Nem él hiába, ki benne marad.
Gyümölcsöt érleljen szereteted,
Tenélküled néped mit se tehet!

3 Jézus világosság, messze ragyog.
Fényéből élnek mind a csillagok.
Bűnünket oszlasd el, nagy a sötét!
Vezesd te jó útra mind, aki vét!

4 Jézus a békesség, fegyvere egy:
A megbocsátó és hű szeretet!
Békére vágyódik mind e világ,
Félelmes ínségből hozzád kiált!

5 Jézus a kezdet és Jézus a vég!
Hadd zengje ég és föld dicséretét!
Vezess az ösvényen, maradj velünk,
Míg örök fényedbe megérkezünk!

English

1 Jesus, Good Shepherd, guide us through the day.
  Lead us with loving grace, show us the way.
  Shelter your flock, O Lord, your people hear.
  Grant us your peaceful rest when night is near.

2 Jesus, the True Vine, we're rooted in you;
  earth us in truth, Lord, then nurture us too.
  Strengthen our faith in distress and in strife;
  we are your branches, in you we have life.

3 Jesus, the Light of all, shine through our world.
  Heavenly stars praise you, radiant Lord.
  Shatter injustice and free us from wrong.
  Light in our darkness, your love is our song.

4 Jesus, our Peace, you're no hawk, but a dove
  granting forgiveness, and healing with love.
  Challenge our apathy, sharpen our will,
  give us the courage to overcome ill.

5 Shepherd and True Vine, our Peace and our Light,
  move through our world, spreading mercy and right.
  Giver of hope and love, bless us always;
  Alpha and Omega, we sing your praise.
  tr. Maggie Hamilton

German

1 Jesus, der gute Hirt, mit dem wir gehen,
  seine Barmherzigkeit, sie lässt uns singen.
  Nimm deine Herd' in acht, wir sind bedroht,
  gib uns die stille Ruh, wenn Nacht sich naht.

2 Jesus, dem Weinstock gleich, hat grosse Kraft,
  dass nicht umsonst gelebt, wer an ihm bleibt.
  Bringen soll viele Frucht die Kraft der Liebe.
  Ohne dich kann dein Volk gar nichts ausrichten.

3 Jesus, das Licht der Welt, strahlt weit hinaus.
  Das Heer der Sterne lebt von seinem Licht.
  Lösch uns're schwere Schuld, das grosse Dunkel.
  Bring wieder auf den Weg, die ihn verfehlten.

4 Jesus, dein Friede hat nur eine Waffe,
  die Liebe, die vergibt und die uns treu bleibt.
  Es sehnt nach Frieden sich uns're Welt,
  sie schreit zu dir den Schrei aus Angst und Not.

5 Weil Jesus Anfang ist, Jesus das Ende,
  singt Erd und Himmel ihm ein ew'ges Lob.
  Führ uns auf deinen Weg und bleibe bei uns,
  bis wir in deinem Licht zu Hause sind.
  tr. Gyula Cseri and Dieter Trautwein

40

Jikelele

Trad. Zulu, South Africa

_____

(I know Jesus, he travels all around the world.)

(Ich kenne Jesus, er reist um die ganze Welt.)

(Je connais Jésus, il voyage à travers le monde entier.)

(Conozco a Jesús. El viaja por todo el mundo.)

41

Jyothi dho Prabhu

Words and music by
Charles Vas, India

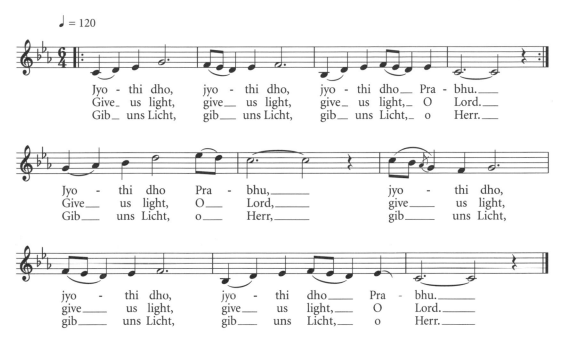

Hindhi

1 Jyothi dho, jyothi dho, jyothi dho Prabhu.
  Jyothi dho, jyothi dho, jyothi dho Prabhu.
  Jyothi dho Prabhu,
  jyothi dho, jyothi dho, jyothi dho Prabhu.

2 Jiivan dho, …   4 Mukthi dho, …

3 Shanthi dho, …   5 Aasish dho, …

English

1 Give us light, give us light, give us light, O Lord.
  Give us light, give us light, give us light, O Lord.
  Give us light, O Lord,
  give us light, give us light, give us light, O Lord.

2 Give us life, …   4 Save us now, …

3 Grant us peace, … 5 Give us grace, …

German

1 Gib uns Licht, gib uns Licht, gib uns Licht, o Herr.
  Gib uns Licht, gib uns Licht, gib uns Licht, o Herr.
  Gib uns Licht, o Herr,
  gib uns Licht, gib uns Licht, gib uns Licht, o Herr.

2 Gib uns Sinn, …   4 Hilf uns nun, …

3 Gib uns Halt, …   5 Gib uns Mut, …

tr. Päivi Jussila

42

Khudaya, rahem kar

R. F. Liberius, Pakistan

♩ = 102

1,3. Khu - da - ya, ra - hem kar.— Khu - da - ya ra -

- hem, khu - da - ya,— ra - hem kar. Khu - da - ya— ra -

Fine

- hem. Khu - da - ya ra - hem kar,— khu - da - ya ra - hem.

2. Ma - si - ha,— ra - hem kar, ma - si - ha,— ra -

- hem. Ma - si - ha ra - hem kar, ma - si - ha,— ra -

D.C.

- hem. Ma - si - ha, ra - hem kar,— ma - si - ha, ra - hem.

(1,3. Lord, have mercy. 2. Christ, have mercy.)

(1,3. Herr, erbarme dich. 2. Christus, erbarme dich.)

(1,3. Seigneur, aie pitié. 2. Christ, aie pitié.)

(1,3. Señor, ten piedad. 2. Cristo, ten piedad.)

Sung unaccompanied

43 | Kyrie eleison

Dinah Reindorf, Ghana

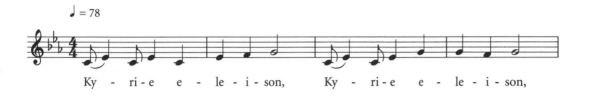

Ky - ri - e e - le - i - son, Ky - ri - e e - le - i - son,

Ky - ri - e e - le - i - son, Ky - ri - e e - le - i - son.

(Lord, have mercy.)

(Herr, erbarme dich.)

(Seigneur, aie pitié de nous.)

(Señor, ten piedad de nosotros.)

44

Kiedy ranne wstają zorze

Franciszek Karpinski   Unknown, Poland

Kie - dy ran - ne   wsta - ją zo - rze,   To - bie zie - mia,   To - bie mo - rze.
In the dis - tance   day is break - ing,   earth and sea a   glad sound mak - ing.
Mor - gen - rö - te   scheint von Fer - ne,   Erd und Meer dich   prei - sen ger - ne,

To - bie śpie - wa   ży - wioł wszel - ki: Bądź po - chwa - lon,   Bo - że wiel - ki!
All cre - a - tion   joined in sing - ing: God, to you, praise   and thanks bring - ing.
al - le Le - be - we - sen sin - gen: Dir, Gott, soll das   Lob - lied klin - gen!

Polish

1 Kiedy ranne wstają zorze,
   Tobie ziemia, Tobie morze.
   Tobie śpiewa żywioł wszelki:
   Bądź pochwalon, Boże wielki!

2 Ledwie oczy przetrzeć zdołam,
   wnet do mego Pana wołam,
   do mego Boga na niebie
   i szukam Go koło siebie.

3 Wielu snem śmierci upadli,
   co się wczoraj spać pokładli,
   my się jeszcze obudzili,
   byśmy Cię, Boże, chwalili.

English

1 In the distance day is breaking,
earth and sea a glad sound making.
All creation joined in singing,
God, to you, praise and thanks bringing.

2 Lord, before you I am standing,
your hand grasping, love adoring,
your own dwelling I will enter,
you, my God, my own life's centre.

3 Friends who walk with us no longer,
held by death, yet love is stronger.
Day dawns bright, new hope extending,
thanks to God in praise unending.
tr. Jeffrey T. Myers

German

1 Morgenröte scheint von Ferne,
Erd und Meer dich preisen gerne,
alle Lebewesen singen:
Dir, Gott, soll das Loblied klingen!

2 Vor dich will ich heute treten,
dich, den Herren, anzubeten,
meinen Gott im Himmel droben
such ich heut und will ihn loben.

3 Viele sind von uns gegangen,
noch hält sie der Tod gefangen.
Weil wir wach den Tag erblicken,
woll'n wir frühes Lob dir schicken.
tr. Tadeusz Sikora

45

Komm, Herr, segne uns

Words and music by
Dieter Trautwein, Germany

Komm, Herr, seg - ne uns, dass wir uns nicht tren - nen,
Bless and keep us, Lord, in your love u - ni - ted,
Viens et nous bé - nis, Dieu de la ren - con - tre,

son - dern ü - ber - all uns zu dir be - ken - nen.
from your fa - mi - ly ne - ver se - par - a - ted.
c'est toi qui u - nis, ton a - mour le mon - tre,

Nie sind wir al - lein, _____ stets sind wir die Dei - nen,
You make all things new _____ as we fol - low af - ter;
gar - de tes en - fants _____ au creux des tem - pê - tes,

La - chen o - der Wei - nen wird ge - seg - net sein.
whe - ther tears or laugh - ter, we be - long to you.
comme au temps des fê - tes sois tou - jours pré - sent.

German

1 Komm, Herr, segne uns,
      dass wir uns nicht trennen,
sondern überall uns zu dir bekennen.
Nie sind wir allein, stets sind wir die Deinen,
Lachen oder Weinen wird gesegnet sein.

2 Keiner kann allein Segen sich bewahren.
Weil du reichlich gibst,
      müssen wir nicht sparen.
Segen kann gedeihn, wo wir alles teilen,
schlimmen Schaden heilen,
      lieben und verzeihn.

3 Frieden gabst du schon,
      Frieden muss noch werden,
wie du ihn versprichst uns zum Wohl auf Erden.
Hilf, dass wir ihn tun, wo wir ihn erspähen;
die mit Tränen säen, werden in ihm ruhn.

4 Komm, Herr, segne uns,
      dass wir uns nicht trennen,
sondern überall uns zu dir bekennen.
Nie sind wir allein, stets sind wir die Deinen,
Lachen oder Weinen wird gesegnet sein.

English

1 Bless and keep us, Lord,
      in your love united,
from your family never separated.
You make all things new as we follow after;
whether tears or laughter, we belong to you.

2 Blessing shrivels up when
      your children hoard it;
help us, Lord, to share, for we can afford it:
blessing only grows in the act of sharing,
in a life of caring, love that heals and glows.

3 Fill your world with peace,
      such as you intended.
Teach us prize the earth, love, replenish, tend it.
Lord, uplift, fulfil all who sow in sadness:
let them reap with gladness, by your kingdom thrilled.

4 Bless and keep us, Lord,
      in your love united,
from your family never separated.
You make all things new as we follow after;
whether tears or laughter, we belong to you.

tr. Fred Kaan

French

1 Viens et nous bénis, Dieu de la rencontre,
c'est toi qui unis, ton amour le montre,
garde tes enfants au creux des tempêtes,
comme au temps des fêtes sois
      toujours présent.

2 Tes dons généreux, ta surabondance
pour les malheureux ouvrent une danse.
Feu qui nous saisit porte témoignage!
Esprit de partage, viens, ô Jésus-Christ.

3 Porteur de la paix, la paix que tu donnes,
semeurs désormais pour le bien des hommes.
Nous voulons, Seigneur, dans la vigilance
que ton règne avance, toi, le vrai bonheur.

4 Viens et nous bénis, Dieu de la rencontre,
c'est toi qui unis, ton amour le montre,
garde tes enfants au creux des tempêtes,
comme au temps des fêtes sois toujours présent.

tr. M. L. Munger and L. Levis

46 | La ténèbre

based on Psalm 139: 12

Jacques Berthier
for the Taizé Community, France

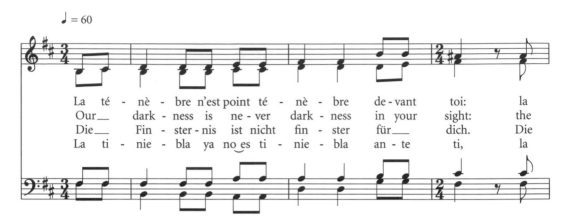

La té - nè - bre n'est point té - nè - bre de - vant toi: la
Our__ dark - ness is ne - ver dark - ness in your sight: the
Die__ Fin - ster - nis ist nicht fin - ster für__ dich. Die
La ti - nie - bla ya no es ti - nie - bla an - te ti, la

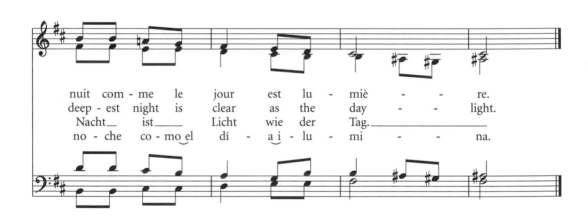

nuit com - me le jour est lu - miè - - re.
deep - est night is clear as the day - - light.
Nacht__ ist__ Licht wie der Tag.__
no - che co - mo el dí - a i - lu - mi - - na.

Lamb of God

Barrie Cabena, Canada

Lamb__ of__ God,__ you take a - way__ the__
A - gneau de__ Dieu,__ qui en - lè - ves__ les__
Got - tes__ Lamm,__ du nimmst hin - weg__ die__

1,2.
sins_____ of the world:____ have mer - cy on us.
pé - chés du__ monde:____ prends pi - tié de nous.
Sün - den der__ Welt:____ er - barme dich un - ser.

3.
world:_____ grant_____ us__ peace.
monde:_____ ac - cor - de - nous__ la__ paix.
Welt:_____ Frie - - den gib_____ uns.

French tr. Marc Chambron; German tr. Dietrich Werner

(Cordero de Dios, que quitas el pecado del mundo: ten piedad de nosotros. Cordero de Dios, que quitas el pecado del mundo: danos tu paz.)

For performance by congregation and/or choir, with organ, or by a cantor with the congregation responding 'have mercy on us' and 'grant us peace'.

48 | Laudate omnes gentes

Psalm 117: 1

Jacques Berthier
for the Taizé Community, France

Lau - da - te om-nes gen - tes, lau - da - te Do-mi - num. Lau -

UPPER VOICE VERSION

(All peoples, praise the Lord!)

(Lobet alle Völker, lobet den Herrn!)

(Louez le Seigneur, tous les peuples!)

(¡Pueblos todos alabad al Señor!)

Le Seigneur fit pour moi des merveilles

Luke 1: 49 (French)

Joseph Gélineau, France

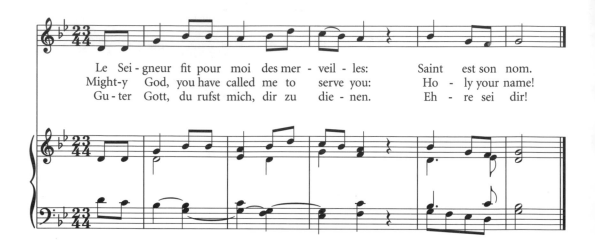

Le Sei - gneur fit pour moi des mer - veil - les: Saint est son nom.
Might-y God, you have called me to serve you: Ho - ly your name!
Gu - ter Gott, du rufst mich, dir zu die - nen. Eh - re sei dir!

English tr. Fred Kaan; German tr. Dietrich Werner

(El Señor ha hecho en mí grandes cosas ¡Santo es su nombre!)

50 Let my prayer rise before you

Psalm 141: 2

David Schack, USA

Let my prayer rise be - fore___ you as in - cense;
Mein Ge - bet stei - ge zu dir, Gott, zum Him - mel.
Ma pri - ère com - me l'en - cens mon - te vers toi,
Su - ba a ti mi o - ra - ción___ co - mo in - cien - so,

the lift - ing up of my hands as the eve - ning sac - ri - fice.
Die Hän - de heb ich zu dir, sag dir Dank für die - sen Tag.
vers toi mes mains se lè - vent en sa - cri - fi - ce le soir.
al - za - das ma - nos sean mi sa - cri - fi - cio de la tarde.

German tr. Dietrich Werner; French tr. Marc Chambron; Spanish tr. Martin Junge

Let there be light

Frances Wheeler Davis

CONCORD 47 76
Robert J. B. Fleming, Canada

Let there be light,_____ let there be un - der - stand - ing,
Licht füll die Welt,_____ den Völ-kern wach - se Hoff - nung,

let all the na - tions gath - er, let them be face to face.
Kul - tu-ren und Na - tion - en rei - chen ein-ander die Hand.

English

1 Let there be light, let there be understanding,
   let all the nations gather, let them be face to face.

2 Open our lips, open our minds to ponder,
   open the door of concord opening into grace.

3 Perish the sword, perish the angry judgement,
   perish the bombs and hunger, perish the fight for gain.

4 Hallow our love, hallow the deaths of martyrs,
   hallow their holy freedom, hallowed be your name.

5 Your kingdom come, your spirit turn to language,
   your people speak together, your spirit never fade.

6 Let there be light, open our hearts to wonder,
   perish the way of terror, hallow the world God made.

German

1 Licht füll die Welt, den Völkern wachse Hoffnung,
   Kulturen und Nationen reichen einander die Hand.

2 Schliesst auf die Tür, die Arm und Reich noch trennen,
   öffnet dem Leben Chancen, Schuldenerlass bringt ein.

3 Hass, den lasst sein! Gebt der Gewalt den Abschied.
   Bomben und Hunger ächtet, Kampf muss nicht länger sein.

4 Liebe gebt Raum, ihre Zeugen achtet!
   Ihre Freiheit schützet! Ihren Kreis macht weit!

5 Gott, dein Reich komm! Dein Geist berühr die Menschen,
   dass Worte Flügel kriegen und Fronten überspring'.

6 Licht füll die Welt, den Herzen wachse Hoffnung,
   Terror und Mord verschwinde, lass endlich Frieden sein.
                    tr. Dietrich Werner

52 Let us break bread together

10 10 with refrain
Trad., African American
arr. Gerre Hancock

English

1 Let us break bread together on our knees.
Let us break bread together on our knees.
When I fall on my knees with my face to the rising sun,
O Lord, have mercy on me.

2 Let us drink wine together on our knees. …

3 Let us praise God together on our knees. …

German

1 Lasst uns Brot brechen und Gott dankbar sein.
Lasst uns Brot brechen und Gott dankbar sein.
Wenn ich knie und hebe den Blick in des Lichtes Schein,
dann, Herr, erbarme dich mein!

2 Nehmt den Kelch, trinkt und lasst uns dankbar sein. …

3 Lasst uns Gott loben und ihm dankbar sein. …
tr. Dieter Trautwein

Spanish

1 De rodillas partamos hoy el pan.
De rodillas partamos hoy el pan.
De rodillas estoy con el rostro al naciente sol.
Oh Dios, apiádate de mi!

2 Compartamos la copa en gratitud. …

3 De rodillas loemos al Señor. …
tr. Federico Pagura

Lobe den Herren

LOBE DEN HERREN 14 14 4 7 8
Words by, and melody adapt. by
Joachim Neander, Germany

Lo - be den Her - ren, den mäch - ti - gen Kö - nig der Eh - ren; lob ihn, o
Praise to the Lord, the Al - migh - ty, the King of cre - a - tion! O my soul,
Peup - les, cri - ez de joie et bon - dis - sez d'al - lé - gres - se. Le Pè - re en -
Al - ma, ben - di - ce al Se - ñor, Dios po - ten - te de glo - ria; de sus mer -

See - le, ver - eint mit den himm - li - schen Chö - ren. Kom - met zu -
praise him, for he is your health and sal - va - tion! Let all who
- voie son Fils ma - ni - fes - ter sa ten - dres - se. Ou - vrons les
- ce - des es - té vi - va en ti la me - mo - ria. Oh des - per -

- hauf: Psal - ter und Har - fe, wacht auf. Las - set den Lob - ge - sang hö - ren.
hear now to his tem - ple draw near, join - ing in glad a - do - ra - tion.
yeux: il est l'i - ma - ge de Dieu! Pour que cha - cun le con - nais - se!
- tad, con voz de go - zo can - tad him - nos de ho - nor y vic - to - ria.

German

1 Lobe den Herren, den mächtigen König der Ehren;
lob ihn, o Seele, vereint mit den himmlischen Chören.
Kommet zuhauf:
Psalter und Harfe, wacht auf.
Lasset den Lobgesang hören.

2 Lobe den Herren, der alles so herrlich regieret,
der dich auf Adelers Fittichen sicher geführet,
der dich erhält,
wie es dir selber gefällt.
Hast du nicht dieses verspüret?

3 Lobe den Herren, der deinen Stand sichtbar gesegnet,
der aus dem Himmel mit Strömen der Liebe geregnet.
Denke daran,
was der Allmächtige kann,
der dir mit Liebe begegnet!

4 Lobe den Herren, was in mir ist, lobe den Namen.
Lob ihn mit allen, die seine Verheissung bekamen.
Er ist dein Licht;
Seele, vergiss es ja nicht.
Lob ihn in Ewigkeit. Amen.

English

1 Praise to the Lord, the Almighty, the King of creation!
O my soul, praise him, for he is your health and salvation!
Let all who hear
now to his temple draw near,
joining in glad adoration.

2 Praise to the Lord, who o'er all things is wondrously reigning,
and, as on wings of an eagle, uplifting, sustaining;
have you not seen
all that is needful has been
sent by his gracious ordaining?

3 Praise to the Lord, who will prosper your work and defend you;
surely his goodness and mercy shall daily attend you;
ponder anew
what the Almighty can do,
if with his love he befriend you.

4 Praise to the Lord! O let all that is in me adore him!
All that has life and breath, come now with praises before him!
Let the amen
sound from his people again:
gladly forever adore him.

tr. Catherine Winkworth (altd.)

(*French and Spanish verses overleaf*)

French

1 Peuples, criez de joie et bondissez d'allégresse.
Le Père envoie son Fils manifester sa tendresse.
Ouvrons les yeux:
il est l'image de Dieu!
Pour que chacun le connaisse!

2 Loué soit notre Dieu, Source et Parole fécondes;
ses mains ont tout créé pour que nos cœurs lui répondent;
par Jésus Christ
il donne l'être et la vie,
en nous sa vie surabonde.

3 Loué soit notre Dieu, dont la splendeur se révèle
quand nous buvons le vin pour une terre nouvelle;
par Jésus Christ
le monde passe aujourd'hui
vers une gloire éternelle.

4 Peuples, battez des mains et proclamez votre fête:
le Père accueille en lui ceux que son Verbe rachète;
dans l'Esprit Saint
par qui vous n'êtes plus qu'un,
que votre joie soit parfaite!
tr. Didier Rimaud

Spanish

1 Alma, bendice al Señor, Dios potente de gloria;
de sus mercedes esté viva en ti la memoria.
Oh despertad,
con voz de gozo cantad
himnos de honor y victoria.

2 Alma, bendice al Señor que a los astros gobierna
y te conduce paciente con mano paterna.
Te perdonó,
de todo mal te libró,
porque su gracia es eterna.

3 Alma, bendice al Señor, de tu vida la fuente,
que te creó y en salud te sostiene clemente;
tu sanador
en todo trance y dolor;
su diestra es omnipotente.

4 Alma, bendice al Señor por su amor infinito;
con todo el pueblo de Dios su alabanza repito;
Dios, mi salud,
de todo bien plenitud,
¡seas por siempre bendito!
tr. Federico Fliedner (altd.)

54 | Louvai ao Senhor

Equipe do CELMU   Simei Pereira do Amaral, Brazil

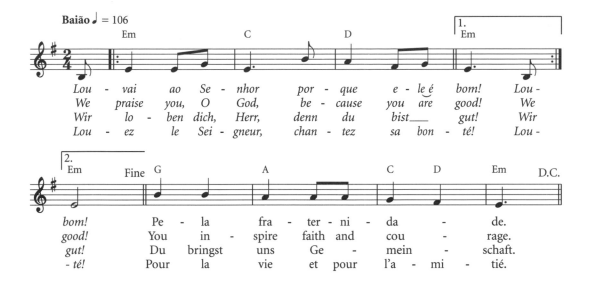

Baião ♩ = 106

Lou - vai   ao   Se - nhor   por - que   e - le é   bom!   Lou -
We   praise you, O   God,   be - cause   you are   good!   We
Wir lo - ben dich, Herr,   denn   du   bist___   gut!   Wir
Lou - ez   le   Sei - gneur, chan - tez   sa   bon - té!   Lou -

bom!   Pe - la   fra - ter - ni - da - de.
good!   You in - spire faith and cou - rage.
gut!   Du   bringst uns   Ge - mein - schaft.
- té!   Pour la   vie   et   pour l'a - mi - tié.

Portuguese

Louvai ao Senhor porque ele é bom!
Louvai ao Senhor porque ele é bom!

1 Pela fraternidade.

2 Pela paz e unidade.

3 Pelo amor e justiça.

4 Pela graça da vida.

English

We praise you, O God, because you are good!
We praise you, O God, because you are good!

1 You inspire faith and courage.

2 You bring nations together.

3 You inspire love and justice.

4 You inspire hope and vision.
                              tr. Fred Kaan

German

Wir loben dich, Herr, denn du bist gut!
Wir loben dich, Herr, denn du bist gut!

1 Du bringst uns Gemeinschaft.

2 Du gibst Segen und Frieden.

3 Du schenkst Einheit und Liebe.

4 Du sorgst für Gerechtigkeit.
                    tr. Ari Kuebelkamp

French

Louez le Seigneur, chantez sa bonté!
Louez le Seigneur, chantez sa bonté!

1 Pour la vie et pour l'amitié.

2 Pour la paix et pour l'unité.

3 Pour la justice et pour l'amour.

4 Pour sa grâce de chaque jour.
                    tr. Marc Chambron

Baião

Instruments: accordion, zabumba, triangle, and other drums

55 | Loving Spirit

Shirley Murray, New Zealand

I-to Loh, Taiwan

Lo - ving Spi - rit, lo - ving Spi - rit,
Geist der Lie - be, Geist der Lie - be,
Es - prit de Dieu ad - mi - ra - ble,

you have cho - sen me to be; you have drawn me
du hast mich ge - ru - fen: Sei! Du hast mich ans
tu m'as choi - si(e), en - voie - moi. Tu m'at - ti - res,

to your won - der, you have set your sign on me.
Licht ge - zo - gen, hast dein Zei - chen ein - ge - setzt.
tu me con - duis vers la beau - té de tes lois.

DRUM

94

English

1 Loving Spirit, loving Spirit,
you have chosen me to be;
you have drawn me to your wonder,
you have set your sign on me.

2 Like a mother, you enfold me,
hold my life within your own,
feed me with your very body,
form me of your flesh and bone.

3 Like a father, you protect me,
teach me the discerning eye,
hoist me up upon your shoulder,
let me see the world from high.

4 Friend and lover, in your closeness
I am known and held and blessed;
in your promise is my comfort,
in your presence I may rest.

5 Loving Spirit, loving Spirit,
you have chosen me to be;
you have drawn me to your wonder,
you have set your sign on me.

German

1 Geist der Liebe, Geist der Liebe,
du hast mich gerufen: Sei!
Du hast mich ans Licht gezogen,
hast dein Zeichen eingesetzt.

2 Mütterlich umarmst du mich,
hältst mich an dir, Tag und Nacht,
sei du selbst mir Kraft und Speise;
ich bin ja dir gleich gemacht.

3 Wie ein Vater schützt du mich,
gibst mir Einsicht und Verstand,
nimmst mich hoch auf deine Schultern,
lässt von dort mich sehn die Welt.

4 Freundschaftlich und voller Liebe
kennst du mich und bist mir nah;
Trost und Segen find ich bei dir,
wo du bist, ist Frieden da.

5 Geist der Liebe, Geist der Liebe,
du hast mich gerufen: Sei!
Du hast mich ans Licht gezogen,
hast dein Zeichen eingesetzt.
tr. Wolfgang Leyk
and Andreas Wöhle

French

1 Esprit de Dieu admirable,
tu m'as choisi(e), envoie-moi.
Tu m'attires, tu me conduis
vers la beauté de tes lois.

2 Comm'une mère tu m'entoures,
tenant ma vie dans tes mains,
tu m'as formé(e) comme ton corps,
tu m'as nourri(e) de ton pain.

3 Comm'un père tu me gardes,
enseigne-moi à aimer
la justice venant de toi,
en tout lieu la contempler.

4 Ami aimé, auprès de toi
je suis chéri(e) et béni(e);
mon réconfort, ma promesse,
c'est en toi que je revis.

5 Esprit de Dieu admirable,
tu m'as choisi(e), envoie-moi.
Tu m'attires, tu me conduis
vers la beauté de tes lois.
tr. Joëlle Gouël

56 | May the love of the Lord

Maria Ling
based on Numbers 6: 24–6

SOON TI
Swee Hong Lim, Singapore

May   the love
Wei_____ yuan
Un - ser Herr

of   the Lord   rest   u-pon   your   soul.
Shen   di   ai   fu - wei ni   di__   ling.
sei   bei dir,   sei - ne Lie - be bei   dir.

May   his love   dwell   in you   through - out ev - ery__
Mei   shi - ke   ta   di ai   zhu   zai ni   xin -
Un - ser Herr   tra - ge dich   durch   den lan - gen__

day.
May his coun - ten - ance shine u - pon you
- li.
Ta di rong - guang zhao - yao__ ni__
Tag.
Got - tes An - ge - sicht mö - ge leuch - ten

and be gra - cious to you.
May his Spir - it
ci - ai ban - sui__ ni.
Wu - lun he - chu
ü - ber dir im - mer - dar.
Sei - ne Hand be -

be__ u - pon you as you leave this__ place.
ta__ di ling dou yong - yuan ban - sui__ ni.
- glei - te dich, nun da du uns ver - lässt.

Mandarin tr. Dong Li; German tr. Angelika Joachim

57

May your breath of love

Maggie Hamilton, England

KUMBA YAH 88 86
Trad., African American/Caribbean

May your breath of love warm our hope, may your
breath of love warm our hope, may your
breath of love warm our hope and heal our bro - ken world.

Gott, dein Lie - bes-hauch wär - me uns, Gott, dein
Lie - bes - hauch wär - me uns, Gott, dein
Lie - bes-hauch wär - me uns und hei - le uns - 're Welt.

Que ton a - mour don - ne l'e - spoir, que ton
a - mour don - ne l'e - spoir, que ton
a - mour don - ne l'e - spoir d'un mon - de qui gué - rit.

Tu gra - tui - to a - mor ál - ze - nos, tu gra -
- tui - to a - mor ál - ze - nos, tu gra -
- tui - to a - mor ál - ze - nos dé al mun - do sa - na - ción.

English

1 May your breath of love warm our hope,
may your breath of love warm our hope,
may your breath of love warm our hope
and heal our broken world.

2 May your kiss of love thrill our hearts, …
and heal our broken world.

3 May your whispered love quell our fears, …
and heal our broken world.

4 May your spark of love fire our dreams, …
to heal our broken world.

5 May your raging love stir our wills, …
to heal our broken world.

6 May your flagrant love fuel our power, …
to heal our broken world.

German

1 Gott, dein Liebeshauch wärme uns,
Gott, dein Liebeshauch wärme uns,
Gott, dein Liebeshauch wärme uns
und heile uns're Welt.

2 Gott, dein Liebeskuss wecke uns, …
und heile uns're Welt.

3 Gott, dein Liebeswort tröste uns, …
und heile uns're Welt.

4 Gott, dein Liebesglanz führe uns, …
und heile uns're Welt.

5 Gott, dein Liebeslicht leite uns, …
und heile uns're Welt.

6 Gott, dein Liebesbrand leuchte uns, …
und heile uns're Welt.
tr. Angelika Joachim
and Päivi Jussila

French

1 Que ton amour donne l'espoir,
que ton amour donne l'espoir,
que ton amour donne l'espoir
d'un monde qui guérit.

2 Que par ton amour dans nos cœurs, …
le monde soit guéri.

3 Viens chasser nos peurs, par amour, …
le monde va guérir.

4 Viens combler nos vœux, par amour, …
le monde va guérir.

5 Ton amour nous pousse en avant, …
le monde va guérir.

6 Ton amour va nous rendre forts, …
le monde va guérir.
tr. Marc Chambron

Spanish

1 Tu gratuito‿amor álzenos,
tu gratuito‿amor álzenos,
tu gratuito‿amor álzenos
dé‿al mundo sanación.

2 Tu tan tierno‿amor tóquenos, …
dé‿al mundo sanación.

3 Tu sereno‿amor cálmenos, …
dé‿al mundo sanación.

4 Tu radiante‿amor guíenos, …
dé‿al mundo sanación.

5 Tu infinito‿amor cólmenos, …
dé‿al mundo sanación.

6 Tu pujante‿amor muévanos, …
dé‿al mundo sanación.
tr. Martin Junge
and Marietta Ruhland

58

Musande Mambo Mwari

Mr Zata, Zimbabwe

From bar 6:

HOSHO (SHAKER)

DRUM

Shona

Leader	Musande	
All	Musande, m'sande, m'sande	
	Mambo Mwari.	
Leader	Musande	
All	Musande, m'sande, m'sande	
	wamasimba.	

1 **Leader** Denga napasi
  All *Mwari wamasimba.*
  Leader zvizere nembiri yenyu.
  All *Mwari wamasimba.*

2 **Leader** Ngaarumbidzwe
  All *Mwari wamasimba.*
  Leader ari kumusorosoro.
  All *Mwari wamasimba.*

3 **Leader** Ngaakudziwe
  All *Mwari wamasimba.*
  Leader samutumwa waMwari.
  All *Mwari wamasimba.*

4 **Leader** Ngaarumbidzwe
  All *Mwari wamasimba.*
  Leader ari kumusorosoro.
  All *Mwari wamasimba.*

  Leader Musande
  All Musande, m'sande, m'sande
    Mambo Mwari.
  Leader Musande
  All Musande, m'sande, m'sande
    wamasimba.

English

  Leader Holy
  All Holy, holy, holy
    Lord God Sabaoth.
  Leader Holy
  All Holy, holy, holy
    God, Almighty One.

1 **Leader** All heaven and earth
  All *Lord God, Almighty One.*
  Leader manifest your glory, O God.
  All *Lord God, Almighty One.*

2 **Leader** Praise be to God.
  All *Lord God, Almighty One.*
  Leader In the highest sing hosanna.
  All *Lord God, Almighty One.*

3 **Leader** Thanks be to God.
  All *Lord God, Almighty One.*
  Leader Blessed is the Saviour who comes,
  All *Lord God, Almighty One.*

4 **Leader** Blessed is He,
  All *Lord God, Almighty One.*
  Leader coming in the name of the Lord.
  All *Lord God, Almighty One.*

  Leader Holy
  All Holy, holy, holy
    Lord God Sabaoth.
  Leader Holy
  All Holy, holy, holy
    God, Almighty One.

tr. Maggie Hamilton

German

  Vorsänger Heilig
  Alle Heilig, heilig, heilig,
    du Herr Zebaoth.
  Vorsänger Heilig
  Alle Heilig, heilig, heilig,
    du Allmächtiger.

1 **Vorsänger** Himmel und Erde
  Alle *Gott, du Allmächtiger.*
  Vorsänger sind voll deiner
    Herrlichkeit, Gott.
  Alle *Gott, du Allmächtiger.*

2 **Vorsänger** Preis sei dir Gott.
  Alle *Gott, du Allmächtiger.*
  Vorsänger Hosianna in der Höhe.
  Alle *Gott, du Allmächtiger.*

3 **Vorsänger** Dank sei dir Gott.
  Alle *Gott, du Allmächtiger.*
  Vorsänger Sei gelobet, du, der da kommt,
  Alle *Gott, du Allmächtiger.*

4 **Vorsänger** Sei gelobet,
  Alle *Gott, du Allmächtiger.*
  Vorsänger der da kommt im Namen
    des Herrn.
  Alle *Gott, du Allmächtiger.*

  Vorsänger Heilig
  Alle Heilig, heilig, heilig,
    du Herr Zebaoth.
  Vorsänger Heilig
  Alle Heilig, heilig, heilig,
    du Allmächtiger.

tr. Päivi Jussila

59

Na nzela na lola

Mwenze Kabemba

Trad., Democratic Republic of Congo

♩ = 120

LEADER

Na nze - la na lo - la to - ko - ta - mbo - la ma - le - mbe, Ma -
We fol - low the Lord a - long the grad - ual road to hea - ven And
So - lan - ge wir ge - hen auf den We - gen uns' - res Got - tes, ist

ALL

- le - mbe to - ko - ta - mbo - la. Ma - le - mbe,___ ma -
know God's king - dom it will come. We know this,___ we
Hoff - nung für die gan - ze Welt. So lasst uns nun auch be -

Ma - le - mbe,___ ma -
We know this,___ we
So lasst uns___ be -

- le - mbe,___ ma - le - mbe to - ko - ta - mbo - la.
know this,___ we know God's king - dom it will come.
- harr - lich___ die We - ge uns' - res Got - tes gehn.

- le - mbe,___
know this,___
- harr - lich___

Lingala

1 **Leader** Na nzela na lola tokotambola malembe,
   All Malembe tokotambola.
   Leader Na nzela na lola tokotambola malembe,
   All Malembe tokotambola.
   All *Malembe, malembe, malembe tokotambola.*
   Malembe, malembe, malembe tokotambola.

2 **Leader** Na nzela na lola tokoyemba na esengo,
   All Malembe tokotambola.

3 **Leader** Na nzela na lola tokoteya na esengo,
   All Malembe tokotambola.

4 **Leader** Na nzela na lola tokobondela masiya,
   All Malembe tokotambola.

English

1 **Leader** We follow the Lord along the gradual road to heaven
   All And know God's kingdom it will come.
   Leader We follow the Lord along the gradual road to heaven
   All And know God's kingdom it will come.
   All *We know this, we know this, we know God's kingdom it will come.*
   We know this, we know this, we know God's kingdom it will come.

2 **Leader** As long as we hope, there is a future for creation,
   All A future for the universe.

3 **Leader** As long as we pray (thank) (sing), there is a future for creation,
   All A future for the universe.

4 **Leader** As long as we act (believe), there is a future for creation,
   All A future for the universe.
       tr. Carolyn Kappauf (altd.)

German

1 **Vorsänger** Solange wir gehen auf den Wegen uns'res Gottes,
   Alle ist Hoffnung für die ganze Welt.
   Vorsänger Solange wir gehen auf den Wegen uns'res Gottes,
   Alle ist Hoffnung für die ganze Welt.
   Alle *So lasst uns nun auch beharrlich die Wege uns'res Gottes gehn.*
   So lasst uns nun auch beharrlich die Wege uns'res Gottes gehn.

2 **Vorsänger** Solange wir hoffen, gibt es Zukunft für die Erde,
   Alle ja Zukunft für die ganze Welt.

3 **Vorsänger** Solange wir beten (danken) (singen), gibt es Zukunft für die Erde,
   Alle ja Zukunft für die ganze Welt.

4 **Vorsänger** Solange wir handeln (glauben), gibt es Zukunft für die Erde,
   Alle ja Zukunft für die ganze Welt.
       tr. Werner Eichel and R. Höcker (altd.)

60

Nú hverfur sól í haf

Sigurbjörn Einarsson

Þorkell Sigurbjörnsson, Iceland

Nú hverf-ur sól_ í haf og húm-ið kem-ur
The sun sets in_ the sea, di-mi-nish-ing_ the
Die Son-ne sinkt ins Meer, rot ist der Dämm'-rung

skjótt. Ég lof-a góð-an Guð, sem gef-ur dag_ og
light. I thank you God of grace, who gives us day_ and
Licht. Mein Lob, Gott, zu_ dir klingt, für Tag und Nacht dir

nótt, minn vök-u-dag, minn draum og nótt.
night, my work-ing day,_ my dream, my night.
singt, für Schla-fen, Wa-chen Dank dir bringt.

Icelandic

1 Nú hverfur sól í haf
og húmið kemur skjótt.
Ég lofa góðan Guð,
sem gefur dag og nótt,
minn vökudag, minn draum og nótt.

2 Þú vakir, faðir vor,
ó, vernda börnin þín,
svo við sem veröld er
og vonarstjarna skín,
ein stjarna hljóð á himni skín.

3 Lát daga nú í nótt
af nýrri von og trú
í myrkri hels og harms
og hvar sem gleymist þú
á jörð, sem átt og elskar þú.

4 Kom, nótt, með náð og frið,
kom nær, minn faðir hár,
og leggðu lyfstein þinn
við lífsins mein og sár,
allt mannsins böl, hvert brot og sár.

English

1 The sun sets in the sea,
diminishing the light.
I thank you God of grace,
who gives us day and night,
my working day, my dream, my night.

2 You are awake, O God,
protecting life that's weak
in all the whole wide world.
The star of hope we seek,
the star of hope of heav'n we seek.

3 As day leads into night
let hope and life break through
where hate and sorrow reign,
where we're forgetting you,
on earth where we're forgetting you.

4 Come night with grace and peace,
be near us, God of love,
come with your healing pow'r
to all the pain of life,
to all the human pain of life.

tr. Per Harling

German

1 Die Sonne sinkt ins Meer,
rot ist der Dämm'rung Licht.
Mein Lob, Gott, zu dir klingt,
für Tag und Nacht dir singt,
für Schlafen, Wachen Dank dir bringt.

2 Du schläfst und schlummerst nicht,
behütest Tag und Nacht,
was klein ist und was schwach.
Lass leuchten dein Angesicht,
lass leuchten uns dein Angesicht.

3 Zu Ende geht der Tag.
Lass Frieden in uns sein.
Wo Hass und Trauer sind,
lass neues Leben sein,
lass neues Leben für alle sein.

4 Bleib bei uns diese Nacht,
sei nah mit deiner Gnad.
Wo Schmerz und Trennung sind,
da lass jetzt Frieden sein;
lass uns, du Heiland, nicht allein.

tr. Dietrich Werner

61
Nu sjunker bullret

Lars Thunberg

T. I. Haapalainen, Finland

Nu sjun - ker bull - ret och stres - sen släp - per.
Nyt me - lu tyyn - tyy, työn pai - ne laan - tuu.
As eve - ning falls, Lord, we lay our bur - dens
Der Lärm ver - ebbt, und die Last wird leich - ter.

Din äng - la - vakt vå - ra mö - dor bär.
En - ke - li kuor - maam - me kos - ket - taa.
in - to your wel - com - ing, car - ing hands.
Es kom - men En - gel und tra - gen mit.

O Gud, väl - sig - na med nat - tens vi - la
Yön le - po siu - naa, oi Her - ra, meil - le.
Bless us with rest in the hours that fol - low,
Gott, seg - ne al - le, die dir ver - trau - en.

det folk vars me - ning du en - sam är!
Si - nul - ta kan - sa - si toi - von saa.
re - new - ing hope - ful - ness through all lands.
Gib Nacht und Ru - he, wo man heut litt.

Swedish

1 Nu sjunker bullret och stressen släpper.
  Din änglavakt våra mödor bär.
  O Gud, välsigna med nattens vila
  det folk vars mening du ensam är!

2 Låt rätten blomma där orätt råder!
  Gör fången fri och bryt våldet ned!
  När natten faller, låt kyrkan vaka
  och mänskan hålla med mänskan fred!

Finnish

1 Nyt melu tyyntyy, työn paine laantuu.
  Enkeli kuormaamme koskettaa.
  Yön lepo siunaa, oi Herra, meille.
  Sinulta kansasi toivon saa.

2 Siis murra kahleet ja väärä valta!
  Suo kukkaan puhjeta oikeuden!
  Suo, että kirkko myös yössä valvoo
  sovinnon liekkinä kaikkien.

tr. T.I. Haapalainen and
Anna-Maija Raittila

English

1 As evening falls, Lord, we lay our burdens
  into your welcoming, caring hands.
  Bless us with rest in the hours that follow,
  renewing hopefulness through all lands.

2 Where there is wrong let your justice flourish;
  banish all violence, set prisoners free!
  And, though we sleep, let our hearts be watchful,
  sharing your vision of unity.

tr. Maggie Hamilton

German

1 Der Lärm verebbt, und die Last wird leichter.
  Es kommen Engel und tragen mit.
  Gott, segne alle, die dir vertrauen.
  Gib Nacht und Ruhe, wo man heut litt.

2 Lass Recht aufblühen, wo Unrecht umgeht.
  Mach die Gefang'nen der Willkür frei.
  Lass deine Kirche mit Jesus wachen
  und Menschen wirken, dass Friede sei.

tr. Jürgen Henkys

62

Nun danket alle Gott

Martin Rinckart

NUN DANKET 67 67 66 66
Johann Crüger

Nun dan-ket al-le Gott mit Her-zen, Mund und Hän-den,
Now thank we all our God, with hearts and hands and voi-ces,
Lou-ons le Cré-a-teur, chan-tons à Dieu lou-an-ges!
De bo-ca y co-ra-zón lo-ad al Dios del cie-lo,

der gros-se Din-ge tut an uns und al-len En-den,
who wond-rous things has done, in whom his world re-joi-ces;
Et joi-gnons no-tre voix au con-cert des saints an-ges!
pues dió-nos ben-di-ción, sa-lud, paz y con-sue-lo.

der uns von Mut-ter-leib und Kin-des-bei-nen an
who, from our moth-er's arms has blessed us on our way
Dès les bras ma-ter-nels il nous a pro-té-gés,
Tan só-lo a su bon-dad de-be-mos nues-tro ser;

un-zäh-lig viel zu gut bis hier-her hat ge-tan.
with count-less gifts of love, and still is ours to-day.
et jus-qu'au der-nier jour, il est no-tre ber-ger.
su san-ta vo-lun-tad nos guí-a por do-quier.

German

1 Nun danket alle Gott
mit Herzen, Mund und Händen,
der grosse Dinge tut
an uns und allen Enden,
der uns von Mutterleib
und Kindesbeinen an
unzählig viel zu gut
bis hierher hat getan.

2 Der ewigreiche Gott
woll uns bei unserm Leben
ein immer fröhlich Herz
und edlen Frieden geben
und uns in seiner Gnad
erhalten fort und fort
und uns aus aller Not
erlösen hier und dort.

3 Lob, Ehr und Preis sei Gott
dem Vater und dem Sohne
und Gott dem Heil'gen Geist
im höchsten Himmelsthrone,
ihm, dem dreiein'gen Gott,
wie es im Anfang war
und ist und bleiben wird
so jetzt und immerdar.

English

1 Now thank we all our God,
with hearts and hands and voices,
who wondrous things has done,
in whom his world rejoices;
who, from our mother's arms
has blessed us on our way
with countless gifts of love,
and still is ours today.

2 O may this bounteous God
through all our life be near us,
with ever joyful hearts
and blessèd peace to cheer us;
and keep us in his grace,
and guide us when perplexed,
and free us from all ills
in this world and the next.

3 All praise and thanks to God
the Father now be given,
the Son, and him who reigns
with them in highest heaven;
the one eternal God,
whom earth and heaven adore;
for thus it was, is now,
and shall be evermore.

tr. Catherine Winkworth

Original version of melody

(French and Spanish verses overleaf)

French

1 Louons le Créateur,
chantons à Dieu louanges!
Et joignons notre voix
au concert des saints anges!
Dès les bras maternels
il nous a protégés,
et jusqu'au dernier jour,
il est notre berger.

2 Loué soit notre Dieu!
Que notre vie entière
tous nous vivions joyeux
sous le regard du Père,
qu'il nous tienne en sa grâce
et nous guide toujours,
nous garde du malheur
par son unique amour.

3 De ce Dieu trois fois saint
qui règne dans la gloire,
chrétiens empressons-nous
de chanter la victoire;
son Royaume est aux cieux
où, plein de majesté,
il règne, seul vrai Dieu,
de toute éternité.

tr. du Pasquier

Spanish

1 De boca y corazón
load al Dios del cielo,
pues diónos bendición,
salud, paz y consuelo.
Tan sólo a su bondad
debemos nuestro ser;
su santa voluntad
nos guía por doquier.

2 Oh Padre celestial,
danos en este día
un corazón filial
y lleno de alegría.
Consérvanos la paz;
tu brazo protector
nos lleve a ver su faz
en tu ciudad, Señor.

3 Dios Padre, Creador,
con gozo te adoramos.
Dios Hijo, Redentor,
tu salvación cantamos.
Dios Santificador,
te honramos en verdad.
Te ensalza nuestra voz,
bendita Trinidad.

tr. Federico Fliedner

63

Nyame ne Sunsum, sian brao!

Caroline Rockson, Ghana

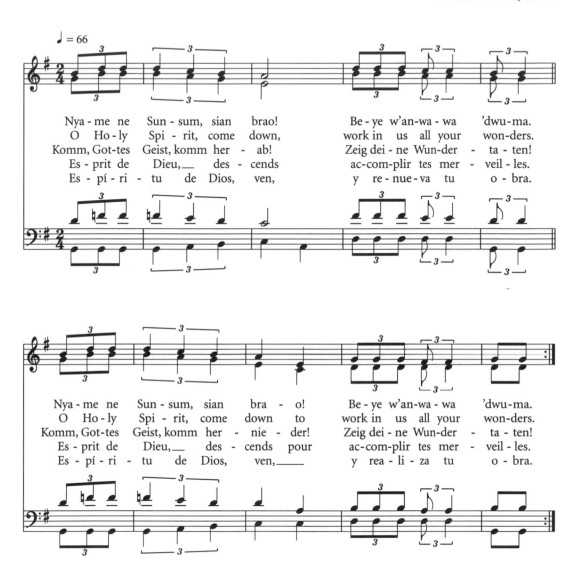

♩ = 66

Nya - me ne Sun - sum, sian brao! Be - ye w'an-wa - wa 'dwu-ma.
O Ho - ly Spi - rit, come down, work in us all your won-ders.
Komm, Got-tes Geist, komm her - ab! Zeig dei - ne Wun-der - ta - ten!
Es - prit de Dieu,__ des - cends ac-com-plir tes mer - veil - les.
Es - pí - ri - tu de Dios, ven, y re - nue-va tu o - bra.

Nya - me ne Sun - sum, sian bra - o! Be - ye w'an-wa - wa 'dwu-ma.
O Ho - ly Spi - rit, come down to work in us all your won-ders.
Komm, Got-tes Geist, komm her - nie - der! Zeig dei - ne Wun-der - ta - ten!
Es - prit de Dieu,__ des - cends pour ac-com-plir tes mer - veil - les.
Es - pí - ri - tu de Dios, ven,__ y rea - li - za tu o - bra.

English tr. Maggie Hamilton; German tr. Dieter Trautwein; French tr. Joëlle Gouël; Spanish tr. Federico Pagura

64

O healing river

Trad., African American

O heal - ing riv - er,_____ send down your
O Fluss des Le - bens,_____ bring dei - ne
Sour - ce qui gué - rit,_____ fais cou - ler
Fuen - te de vi - da,_____ en - vía las

wa - ters,_____ send down your wa - ters_____ up - on this
Flu - ten,_____ lass Was - ser strö - men_____ in die - ses
tes flots,_____ fais cou - ler tes flots_____ sur cet - te
a - guas,_____ en - vía las a - guas_____ so - bre la

land._____ O heal - ing riv - er,_____ send down your
Land._____ O Fluss des Le - bens,_____ dein Was - ser
ter - re._____ Sour - ce qui gué - rit,_____ fais cou - ler
tier - ra._____ Fuen - te de vi - da,_____ en - vía las

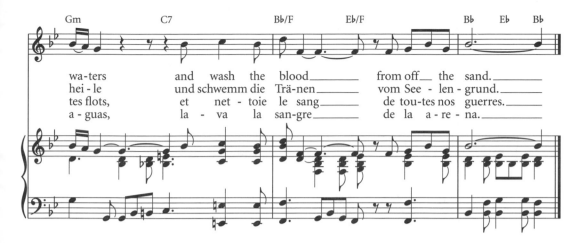

wa-ters and wash the blood_____ from off__ the sand._____
hei-le und schwemm die Trä-nen_____ vom See-len-grund._____
tes flots, et net-toie le sang_____ de tou-tes nos guerres._____
a-guas, la-va la san-gre_____ de la a-re-na._____

English

1 O healing river, send down your waters,
   send down your waters upon this land.
   O healing river, send down your waters
   and wash the blood from off the sand.

2 This land is parching, this land is burning,
   no seed is growing in the barren ground.
   O healing river, send down your waters.
   O healing river, send your waters down.

3 Let the seed of freedom awake and flourish.
   Let the deep roots nourish, let the tall stalks rise.
   O healing river, send down your waters;
   O healing river from out of the skies.

German

1 O Fluss des Lebens, bring deine Fluten,
   lass Wasser strömen in dieses Land.
   O Fluss des Lebens, dein Wasser heile
   und schwemm die Tränen vom Seelengrund.

2 Das Land vertrocknet, versengt, verdurstet,
   kein grüner Grashalm auf dürrem Grund.
   O Fluss des Lebens, bring deine Fluten,
   dein Wasser fliesse, o Lebensstrom.

3 Die Saat der Freiheit lass keimen, wachsen,
   und tief verwurzelt lass sie erblühn.
   O Fluss des Lebens, mach neu die Erde,
   dein Wasser ströme aus Himmelshöh'n.
                              tr. Horst Bracks

French

1 Source qui guérit, fais couler tes flots,
   fais couler tes flots sur cette terre.
   Source qui guérit, fais couler tes flots,
   et nettoie le sang de toutes nos guerres.

2 Dure est notre terre où tout est brûlé,
   pas un grain ne pousse sur son sol durci.
   Source qui guérit, fais couler tes flots,
   fais couler tes flots, source qui guérit.

3 Que la liberté puisse enfin fleurir
   pour que ses racines soient plus profondes.
   Source qui guérit, fais couler tes flots,
   que tombe la pluie du ciel sur le monde.
                              tr. Marc Chambron

Spanish

1 Fuente de vida, envía las aguas,
   envía las aguas sobre la tierra.
   Fuente de vida, envía las aguas
   lava la sangre de la arena.

2 La tierra gime, la tierra arde,
   nada florece en tal hoguera.
   Fuente de vida, envía las aguas
   fuente eterna, danos la vida.

3 La libertad, oh, haz germinar,
   profundas raíces, produzcan flores.
   Fuente de vida, envía las aguas
   fuente eterna, abre los cielos.
                              tr. Martin Junge

113

65

O God, we call

O GOD WE CALL 86 88
Words and music by
Linnea Good, Canada

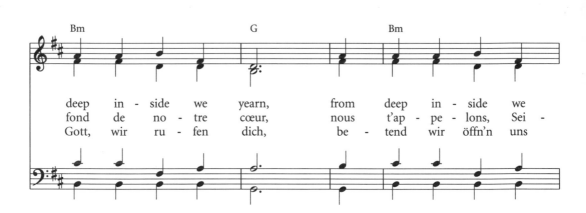

O God, we call, O God, we call, from
O toi, Seigneur, nous t'appelons, du
Zu dir wir flehn, zu dir uns sehn', O

deep inside we yearn, from deep inside we
fond de notre cœur, nous t'appelons, Sei-
Gott, wir rufen dich, betend wir öffn'n uns

yearn, from deep inside we yearn for you.
-gneur, oui, nous te désirons, Seigneur.
dir, alles in uns sehnt sich nach dir.

French tr. Marc Chambron; German tr. Dietrich Werner

66

O let us spread the pollen of peace (The pollen of peace)

Words and music by
Roger Courtney, Northern Ireland

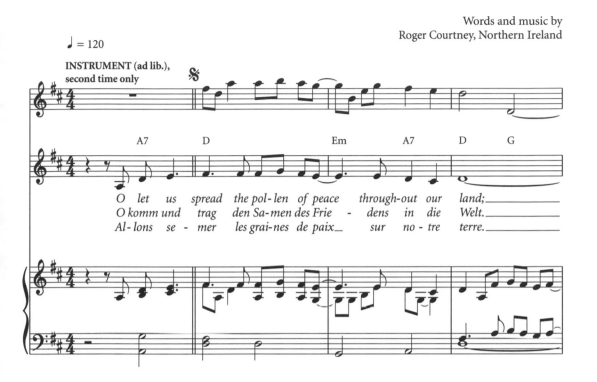

O let us spread the pol-len of peace through-out our land;_____
O komm und trag den Sa-men des Frie - dens in die Welt._____
Al-lons se - mer les grai-nes de paix_ sur no - tre terre._____

— let us spread the pol-len of peace through-out our land;_____ let us
— Mit uns trag den Sa-men des Frie - dens in die Welt._____ Mit uns
— Ré-pan-dons les grai-nes de paix_ sur no - tre terre._____ Ré-pan-

spread   the pol-len of peace_   and make all con-flicts cease,   let us
trag   den Sa-men des Frie   -   dens wo Un-frie-de herrscht.   Mit uns
- dons   les grai-nes de paix,_   et ar-rê-tons les guerres.   Ré-pan-

Fine

spread   the pol-len of peace_   through-out our   land._____
trag   den Sa-men des Frie   -   dens in die Welt._____
- dons   les grai-nes de paix_   sur no-tre terre._____

O Christ   has sown the seeds of love;_____   O
O Herr,   du schenkst Ver-söh-nung uns._____   Du
Christ a   se-mé par-tout l'a-mour._____   Sa

German tr. Angelika Joachim and Päivi Jussila; French tr. Marc Chambron

67

O living water, refresh my soul

Words and music by
Rosalie Vissing, USA

INSTRUMENT (ad lib.)

O liv-ing wa - ter, re-fresh my soul. O liv-ing
Was - ser des Le - bens, er - qui-cke mich. Was - ser des
Toi, l'eau vi - van - te, bai-gne mon âme. Toi, l'eau vi -

wa - ter, re-fresh my soul. Spi - rit of joy,
Le - bens, er - qui-cke mich. Du Geist der Kraft,
- van - te, ra - fraî-chis - moi. Es - prit de joie,

Lord of cre - a - tion, Spi - rit of hope, Spi - rit of peace.
du Gott der Schöp - fung. Du Geist der Zeit, du Geist der Ruh.
ô Dieu cré - a - teur, Es - prit d'es-poir, Es - prit de paix.

Fine

Spi - rit of God.   Spi - rit of God.
Du Heil - 'ger Geist.   Du Heil - 'ger Geist.
Es - prit de Dieu.   Es - prit de Dieu.

English

O living water, refresh my soul.
O living water, refresh my soul.
Spirit of joy, Lord of creation,
Spirit of hope, Spirit of peace.

1 Spirit of God. Spirit of God.

2 O set us free. O set us free.

3 Come, pray in us. Come, pray in us.

German

Wasser des Lebens, erquicke mich.
Wasser des Lebens, erquicke mich.
Du Geist der Kraft, du Gott der Schöpfung.
Du Geist der Zeit, du Geist der Ruh.

1 Du Heil'ger Geist. Du Heil'ger Geist.

2 Befreie uns. Befreie uns.

3 Bete in uns. Bete in uns.
tr. Angelika Joachim
and Päivi Jussila

French

Toi, l'eau vivante, baigne mon âme.
Toi, l'eau vivante, rafraîchis-moi.
Esprit de joie, ô Dieu créateur,
Esprit d'espoir, Esprit de paix.

1 Esprit de Dieu. Esprit de Dieu.

2 Délivre-nous. Délivre-nous.

3 Tu pries en nous. Tu pries en nous.
tr. Marc Chambron

68 | O Saint Esprit, viens sur nous

Bayiga Bayiga, Cameroon

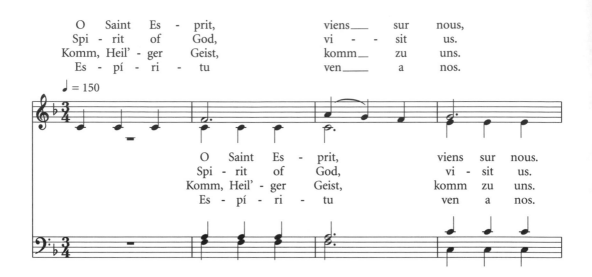

O Saint Es - prit, viens__ sur nous,
Spi - rit of God, vi - sit us.
Komm, Heil' - ger Geist, komm__ zu uns.
Es - pí - ri - tu ven__ a nos.

♩ = 150

O Saint Es - prit, viens sur nous.
Spi - rit of God, vi - sit us.
Komm, Heil' - ger Geist, komm zu uns.
Es - pí - ri - tu ven a nos.

Com - ble - nous de tes__ fa - veurs.
With your bless - ings fill__ our lives.
Und dein Se - gen fül - le uns.
Llé - na - nos de tu__ gra - cia.

English tr. Maggie Hamilton; German tr. Dietrich Werner; Spanish tr. Dorli Schwab

69

Oré poriajú verekó

Trad. Guarani, Paraguay

Guarania ♩ = 72

1. O - ré po - ria - jú ve - re - kó, Ñan - de -
2. O - ré po - ria - jú ve - re - kó, Je - su -

- ya - ra. O - ré po - ria - jú ve - re - kó, Ñan - de - ya - ra.
- cris - to. O - ré po - ria - jú ve - re - kó, Je - su - cris - to.

(Lord, have mercy. Christ, have mercy.)

(Herr, erbarme dich. Jesus Christus, erbarme dich.)

(Seigneur, aie pitié de nous. Jésus-Christ, aie pitié.)

(Señor, ten piedad. Jesucristo, ten piedad.)

Guarania

70
Ososŏ

based on 2 Corinthians 5: 17–20

Words and music by
Geonyong Lee, Korea

O - so - sŏ___ o - so - sŏ, pyŏng-hwa - ŭi___ im - gŭm
Come, now, O___ Prince of peace, make us one___ bo - dy.
Komm nun, Fürst des Frie - dens, ein uns zu dei - nem Lei - be,
Viens, viens, ô___ roi de paix et ras - sem - ble - nous,
Ven, Dios, Prín - ci - pe de paz, haz - nos un___ cuer - po;

u - ri - ga___ han - mom i - ru - ge ha - so - sŏ.
Come, O Lord_ Je - sus, re - con - cile your_ peo - ple.
komm, Herr Je - sus, komm, ver - söh - ne dir all___ dein Volk.
viens, viens, ô___ maî - tre, ré - con - ci - lie ton peu - ple.
ven, Se - ñor,___ Je - sús, y re - con - ci - lia a tu pue - blo.

Korean

1 Ososŏ ososŏ,
   pyŏnghwaŭi imgŭm
   uriga hanmom
   iruge hasosŏ.

2 Ososŏ ososŏ,
   sarangŭi imgŭm,
   uriga hanmom
   iruge hasosŏ.

3 Ososŏ ososŏ,
   chayuŭi imgŭm,
   uriga hanmom
   iruge hasosŏ.

4 Ososŏ ososŏ,
   tongilŭi imgŭm,
   uriga hanmom
   iruge hasosŏ.

English

1 Come, now, O Prince of peace,
make us one body.
Come, O Lord Jesus,
reconcile your people.

2 Come, now, O God of love,
make us one body.
Come, O Lord Jesus,
reconcile your people.

3 Come now and set us free,
O God, our Saviour.
Come, O Lord Jesus,
reconcile all nations.

4 Come, Hope of unity,
make us one body.
Come, O Lord Jesus,
reconcile all nations.

tr. Marion Pope

German

1 Komm nun, Fürst des Friedens,
ein uns zu deinem Leibe,
komm, Herr Jesus, komm,
versöhne dir all dein Volk.

2 Komm, nun, Gott der Liebe,
ein uns zu deinem Leibe,
komm, Herr Jesus, komm,
versöhne dir all dein Volk.

3 Komm, nun und befrei uns,
denn du bist unser Retter,
komm, Herr Jesus, komm,
versöhne dir alle Völker.

4 Komm, erhoffte Einheit,
ein uns zu einem Leibe,
komm, Herr Jesus, komm,
versöhne dir alle Völker.

tr. Dieter Trautwein
and Dietrich Werner

French

1 Viens, viens, ô roi de paix
et rassemble-nous,
viens, viens, ô maître,
réconcilie ton peuple.

2 Viens, viens, ô roi d'amour
et rassemble-nous,
viens, viens, ô maître,
réconcilie ton peuple.

3 Viens, viens, libérateur
et libère-nous,
viens, viens, ô maître,
réconcilie ton peuple.

4 Viens, viens, source d'espoir
fais de nous ton corps,
viens, viens, ô maître,
réconcilie les peuples.

tr. Robert Faerber

Spanish

1 Ven, Dios, Príncipe de paz,
haznos un cuerpo;
ven, Señor, Jesús,
y‿reconcilia‿a tu pueblo.

2 Ven, Dios, Príncipe de‿Amor,
haznos sólo‿un cuerpo;
ven, Señor Jesús,
y‿reconcilia‿a tu pueblo.

3 Ven, Señor, liberanos
Oh, Dios, Salvador;
ven, Señor Jesús,
y‿reconcilia‿a tu pueblo.

4 Fuente de‿esperanza, Dios,
haznos sólo‿un cuerpo;
ven, Señor Jesús,
y‿reconcilia‿a tu pueblo.

tr. Martin Junge

71 | Ouve, Senhor, eu estou clamando (Tem piedade)

based on Psalm 102: 1–2

Unknown, Brazil

Ou - ve, Se - nhor,_____ eu es - tou cla - man - do,_____
Hear me, O Lord,_____ O__ hear my cry - ing;__
Hö - re mich, Herr,_____ wenn ich zu dir ru - fe.__
Pi - tié, Sei - gneur,_____ c'est vers toi que je crie,__

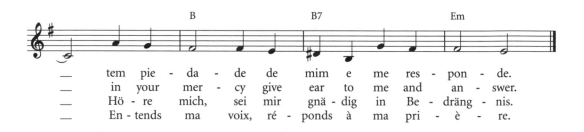

— tem pie - da - de de mim e me res - pon - de.
— in your mer - cy give ear to me and an - swer.
— Hö - re mich, sei mir gnä - dig in Be - dräng - nis.
— En - tends ma voix, ré - ponds à ma pri - è - re.

English tr. Alan Luff; German tr. Päivi Jussila

124

72 Porque él entró (Tenemos esperanza)

Federico J. Pagura, Argentina

Homero Perera, Uruguay

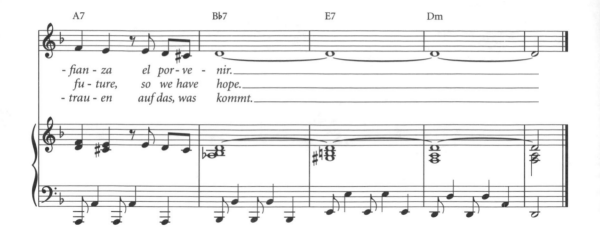

- fian - za   el por - ve - nir._____
fu - ture,   so we have hope._____
- trau - en auf das, was kommt._____

Spanish

1 Porque él entró en el mundo y en la historia;
porque él quebró el silencio y la agonía;
porque llenó la tierra de su gloria;
porque fue luz en nuestra noche fría;
porque él nació en un pesebre oscuro;
porque él vivió sembrando amor y vida;
porque partió los corazones duros
y levantó las almas abatidas.
Por eso es que hoy tenemos esperanza;
por eso es que hoy luchamos con porfía;
por eso es que hoy miramos con confianza
el porvenir en esta tierra mía.
Por eso es que hoy tenemos esperanza;
por eso es que hoy luchamos con porfía;
por eso es que hoy miramos con confianza
el porvenir.

2 Porque atacó a ambiciosos mercaderes
y denunció maldad e hipocresía;
porque exaltó a los niños, las mujeres,
y rechazó a los que de orgullo ardían.
Porque él cargó la cruz de nuestras penas
y saboreó la hiel de nuestros males;
porque aceptó sufrir nuestra condena
y así morir por todos los mortales.

3 Porque una aurora vió su gran victoria
sobre la muerte, el miedo, las mentiras;
ya nada puede detener su historia,
ni de su Reino eterno la venida.

Tango ensemble: piano, bandonéon (accordion), guitars

English

1 Because he came into our world and story,
   because he heard our silence and our sorrow;
   because he filled the whole world
           with his glory,
   and came to light the darkness
           of our morrow;
   because his birth was in a darkened corner,
   because he lived proclaiming life and love;
   because he quickened hearts that
           had been dormant,
   and lifted those whose lives had
           been down-trodden:
   so we today have hope and expectation,
   so we today can struggle with conviction,
   so we today can trust we have a future,
   so we have hope in this our world of tears.
   So we today have hope and expectation,
   so we today can struggle with conviction,
   so we today can trust we have a future,
   so we have hope.

2 Because he drove the merchants
           from the temple,
   denouncing evil and hypocrisy;
   because he raised up little ones and women,
   and put down all the mighty from their seats;
   because he bore the cross for our wrongdoings,
   and understood our failings and our weakness;
   because he suffered from our condemnation
   and then he died for every mortal creature:

3 Because of victory one morning early,
   when he defeated death and fear and sorrow,
   so nothing can hold back his mighty story
   nor his eternal kingdom tomorrow:
           tr. Len Lythgoe

German

1 In unser Leben, in uns're Geschichte
   ist er gekommen, sie mit uns zu teilen,
   und hat das Schweigen,
           hat die Angst zerbrochen
   und uns're Dunkelheit
           in Licht verwandelt.
   In einem unbekannten Stall geboren,
   um Liebe auszusäen und neues Leben,
   erstarrte Herzen endlich zu erweichen
   und die zu stützen, die am Boden liegen.
   Und darum sind wir heute voller Hoffnung,
   und darum kämpfen wir heut' ohne Zittern,
   und darum blicken wir heut' voll Vertrauen,
   in eine neue Zukunft für uns alle.
   Und darum sind wir heute voller Hoffnung,
   und darum kämpfen wir heut' ohne Zittern,
   und darum blicken wir heut' voll Vertrauen
   auf das, was kommt.

2 Gegen den Ehrgeiz der Geschäftemacher
   hat er gekämpft, und gegen jede Lüge,
   den Frauen, Kindern eig'nen Wert gegeben,
   aber die stolz und hart sind abgewiesen.
   Er trug mit uns das Kreuz all uns'rer Schmerzen
   und litt wie wir die Qual all uns'rer Übel,
   war selbst bereit, der Menschen Schuld zu teilen,
   um so den Tod für immer zu besiegen.

3 Weil uns're Welt dies Zeichen seiner Macht sah
   über den Tod, die Angst und alle Lügen,
   ist heute schon sein Wirken unaufhaltsam
   und wird auf Dauer niemals unterliegen.
           tr. Claudia Lohff Blatezky

73

Praise the God of all creation

Words and music by
Marty Haugen, USA

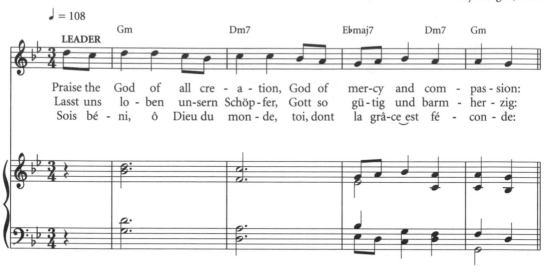

English

1 **Leader** Praise the God of all creation, God of mercy and compassion:
   All *Alleluia! Alleluia! Praise the Word of truth and life!*

2 Tree of life and endless wisdom, be our root, our growth and glory:

3 Living water, we are thirsting for the life that you have promised:

4 Come, O Spirit, kindle fire in the hearts of all your people:

5 Praise the God of all creation, God of mercy and compassion:

German

1 **Vorsänger** Lasst uns loben unsern Schöpfer, Gott so gütig und barmherzig:
   Alle *Halleluja! Halleluja! Gottes Wort schenkt Lebenskraft.*

2 Lebensbaum, wir wollen wachsen, in Gott wurzeln und vertrauen:

3 Gott, wir dürsten nach dem Wasser, das uns Leben neu verheisst:

4 Komm, erfüll uns, Geist des Lebens, mit dem Feuer deiner Vollmacht:

5 Lasst uns loben unsern Schöpfer, Gott so gütig und barmherzig:
   tr. Dorothea Wulfhorst

French

1 **Solo** Sois béni, ô Dieu du monde, toi, dont la grâce est féconde:
   Tous *Alléluia! Alléluia! Ta Parole est notre joie.*

2 Tu donnes vie et sagesse, sois notre force sans cesse:

3 Eau vivante, jaillissante, en nous source bienfaissante:

4 Viens habiter, par l'Esprit Saint, dans le cœur de tous les humains:

5 Sois béni, ô Dieu du monde, toi, dont la grâce est féconde:
   tr. Marc Chambron

74

Praise the One who breaks the darkness

Rusty Edwards

NETTLETON 87 87D
Unknown, USA

Praise the One who breaks the dark-ness with a li - ber - a - ting light;
Lob sei dir: All un - ser Dun - kel machst du hell mit dei-nem Licht.

praise the One who frees the pris - 'ners, turn-ing blind - ness in - to sight.
Lob sei dir: Wer tief ge - fang-en, wer blind, er - lan - get ne - ue Sicht.

Praise the_ One who preached the gos - pel, heal-ing_ ev - 'ry dread dis - ease,
Lob sei_ dir: die fro - he Bot - schaft, tie - fe_ Hei - lung du uns bracht,

calm-ing storms and feed-ing thou-sands with the ve - ry bread of peace.
gros - se Stür - me und den Hun - ger viel - er hast du still ge - macht.

English

1 Praise the One who breaks the darkness with a liberating light;
praise the One who frees the pris'ners, turning blindness into sight.
Praise the One who preached the gospel, healing ev'ry dread disease,
calming storms and feeding thousands with the very bread of peace.

2 Praise the One who blessed the children with a strong yet gentle word;
praise the One who drove out demons with a piercing, two-edged sword.
Praise the One who brings cool water to the desert's burning sand;
from this well comes living water quenching thirst in ev'ry land.

3 Praise the one true love incarnate: Christ, who suffered in our place.
Jesus died and rose for many that we may know God by grace.
Let us sing for joy and gladness, seeing what our God has done.
Praise the one redeeming glory; praise the One who makes us one.

German

1 Lob sei dir: All unser Dunkel machst du hell mit deinem Licht.
Lob sei dir: Wer tief gefangen, wer blind, erlanget neue Sicht.
Lob sei dir: Die frohe Botschaft, tiefe Heilung du uns bracht,
grosse Stürme und den Hunger vieler hast du still gemacht.

2 Lob sei dir: Den kleinen Kindern galt dein zartes Segenswort.
Lob sei dir: Die dunklen Mächte durch dein Machtwort ziehen fort.
Lob sei dir: Die trocknen Wüsten lässt du ergrünen zu voller Pracht.
Durst nach Leben findet Antwort in dem Leben, das du schenkst.

3 Lob sei dir, du wahre Liebe: Christus, der für uns gelitten.
Jesus starb und ist erstanden, dass wir neu erkennen Gott.
Lasst uns jubeln, lasst uns singen, spüren, was Gott in uns tut.
Lob sei dir: du schenkst uns Freiheit. Lob sei dir, du machst uns eins.
tr. Dietrich Werner

75

Put peace into each other's hands

Fred Kaan, Netherlands/England

Ron Klusmeier, Canada

Put peace___ in - to___ each oth - er's
Legt Frie - den ein - an - der in___ die

hands___ and like___ a trea - sure hold
Hand,___ und wie___ ein Schatz___ be - schützt

it, pro - tect it like___ a can - dle
ihn, be - wahrt ihn wie___ ein Ker - zen -

flame, with ten - der - ness en - fold it.___
- licht, mit Zärt - lich - keit ihn ent - fal - tet.___

English

1 Put peace into each other's hands
and like a treasure hold it,
protect it like a candleflame,
with tenderness enfold it.

2 Put peace into each other's hands
with loving expectation;
be gentle in your words and ways,
in touch with God's creation.

3 Put peace into each other's hands
like bread we break for sharing;
look people warmly in the eye:
our life is meant for caring.

4 Give thanks for strong—yet tender—hands,
held out in trust and blessing.
Where words fall short, let hands speak out,
the heights of love expressing.

5 Reach out in friendship, stay with faith
in touch with those around you.
Put peace into each other's hands:
the Peace that sought and found you.

Alternative (original) verses 4 and 5

4 As at communion, shape your hands
into a waiting cradle;
the gift of Christ receive, revere,
united round the table.

5 Put Christ into each other's hands,
he is love's deepest measure;
in love make peace, give peace a chance,
and share it like a treasure.

German

1 Legt Frieden einander in die Hand,
und wie ein Schatz beschützt ihn,
bewahrt ihn wie ein Kerzenlicht,
mit Zärtlichkeit ihn entfaltet.

2 Legt Frieden einander in die Hand,
voll Hoffnung und Erwartung,
gewaltfrei sei euer Wort und Werk,
es achte Gottes Schöpfung.

3 Legt Frieden einander in die Hand,
wie Brot, das einander wir teilen.
Seht Menschen freundlich ins Angesicht,
begegnet einander in Liebe.

4 Dank sei für jede off'ne Hand
entgegengestreckt dem Fremden.
Wo Worte fehlen, lasst Hände sprechen,
ein Zeichen setzen für Frieden.

5 In Freundschaft bleibt euch zugewandt,
im Glauben fest verbunden.
Legt Frieden einander in die Hand,
den Frieden, den Gott lässt wachsen.
tr. Dietrich Werner

76

Rakanaka Vhangeri

Trad. Shona, Zimbabwe

Shona

1 Rakanaka Vhangeri, rakanaka.
Rakanaka Vhangeri, rakanaka.
Rakanaka Vhangeri, rakanaka.
Rakanaka Vhangeri, rakanaka.
Nda nguri ndakuudza kuti rakanaka.
Nda nguri ndakuudza kuti rakanaka.
Nda nguri ndakuudza kuti rakanaka.
Nda nguri ndakuudza kuti rakanaka.

2 Rinesimba Vhangeri, rinesimba.
Rinesimba Vhangeri, rinesimba.
Nda nguri ndakuudza kuti rinesimba. …

3 Ngariende Vhangeri, ngariende.
Ngariende Vhangeri, ngariende.
Nda nguri ndakuudza kuti ngariende. …

English

1 Come and hear now the gospel, hear the good news.
Come and hear now the gospel, hear the good news.
Come and hear now the gospel, hear the good news.
Come and hear now the gospel, hear the good news.
We sing it many times again, the gospel's good news.
We sing it many times again, the gospel's good news.
We sing it many times again, the gospel's good news.
We sing it many times again, the gospel's good news.

2 Powerful is the gospel, very powerful.
Powerful is the gospel, very powerful.
We sing it many times again, the gospel's powerful. …

3 Let us live out the gospel, live the good news.
Let us live out the gospel, live the good news.
We sing it many times again, go, live the good news. …
tr. Francis Brienan and Maggie Hamilton

German

1 Gottes Wort tut uns gut! Ja, es tut gut!
Gottes Wort tut uns allen gut! Ja, es tut gut!
Gottes Wort tut uns gut! Ja, es tut gut!
Gottes Wort tut uns allen gut! Ja, es tut gut!
Ich habe dir schon immerzu erzählt: Es tut gut!
Ich habe dir schon immerzu erzählt: Es tut gut!
Ich habe dir schon immerzu erzählt: Es tut gut!
Ich habe dir schon immerzu erzählt: Es tut gut!

2 Gottes Wort gibt uns Kraft! Ja, es gibt Kraft!
Gottes Wort gibt uns allen Kraft! Ja, es gibt Kraft!
Ich habe dir schon immerzu erzählt: Es gibt Kraft! …

3 Gottes Wort ruft nach Boten! Ja, es ruft uns!
Gottes Wort ruft nach vielen Boten! Ja, es ruft uns!
Ich habe dir schon immerzu erzählt: Es ruft uns! …
tr. Hendriks Mavunduse and Dieter Trautwein

77

Reamo leboga

Trad., Botswana

♩. = 68

Re - a - mo le - - bo - ga, re - a - mo
We give our thanks_____ to God, we give our
Wir dan - ken un - - serm Gott, wir dan - ken
Mer - ci à toi,_____ Sei - gneur, mer - ci à

le - bo - ga, re - a - mo le - bo - ga mo - di - mo wa ro - na.
thanks to God, we give our thanks to God, we give thanks to our God.
un - serm Gott, wir dan - ken un - serm Gott, der Dank gilt un - serm Gott.
toi, Sei - gneur, mer - ci à toi, Sei - gneur, nous te ren - dons grâ - ce.

2. Ga a yo yo tshwa-nang le - we - na
3. Re - pho - lo si - tswe ke - we - na

siTswana

1 Reamo leboga,
 reamo leboga,
 reamo leboga
 modimo wa rona.

2 Ga a yo yo tshwanang lewena, …
 modimo wa rona.

3 Repholo sitswe kewena, …
 modimo wa rona.

English

1 We give our thanks to God,
 we give our thanks to God,
 we give our thanks to God,
 we give thanks to our God.

2 There is no-one like God, …
 there's no-one like our God.

3 We have been saved by God, …
 we've been saved by our God.
 tr. I-to Loh

German

1 Wir danken unserm Gott,
 wir danken unserm Gott,
 wir danken unserm Gott,
 der Dank gilt unserm Gott.

2 Denn Gott ist niemand gleich, …
 niemand gleicht unserm Gott.

3 Wir sind durch Gott befreit, …
 befreit durch unsern Gott.
 tr. Wolfgang Leyk

French

1 Merci à toi, Seigneur,
 merci à toi, Seigneur,
 merci à toi, Seigneur,
 nous te rendons grâce.

2 Aucun n'est comme toi, …
 comme toi, ô Seigneur.

3 C'est lui qui nous sauva. …
 C'est Dieu qui nous sauva.
 tr. Joëlle Gouël

78

Sa harap ng pagkaing dulot mo

Rolando S. Tinio
based on Acts 14:17

LATOK
Francisco F. Feliciano, Philippines

Sa ha-rap ng pag - ka - ing du-lot mo_____ sa____ a - min
With the grace of this day's food, that you lay_____ be - fore us,
Das__ Mahl, uns be - rei-tet: ein__ Zei - chen dei-ner Gna - de,

wa-lang sa - wa ka - ming__ nag - pa - pa-sa - la - mat.
bless us still_____ as we thank you with hearts____ o - ver - flow - ing.
un-sern Dank wir vor dich brin-gen, ein Se - gens-lied dir sin-gen,

Ang i - yong_____ pat - nu - bay hi - ni - hi - ngi__ na - min,
And be - stow_____ on__ us,__ Lord, the guid - ance that we yearn for,
führ und lei - te du uns Gott, dein Licht er - hel - le die-se Ta - ge,

140

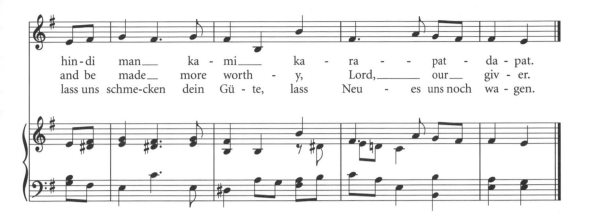

Tagalog

1 Sa harap ng pagkaing dulot mo sa amin
walang sawa kaming nagpapasalamat.
Ang iyong patnubay hinihingi namin,
hindi man kami karapatdapat.

2 Walang sawa sa amin, turo mong pag-asa
kaya't sana'y diggin ang aming dalangin.
Ang pagkaing dulot balutin ng grasya,
nang ang buhay ko'y 'yong marapatin.

English

1 With the grace of this day's food, that you lay before us,
bless us still as we thank you with hearts overflowing.
And bestow on us, Lord, the guidance that we yearn for,
and be made more worthy, Lord, our giver.

2 You have never abandoned those who seek your counsel;
so we pray that this meal be a meal for the spirit.
May we grow in wisdom with ev'ry day that passes,
and be made deserving of your bounty.
tr. Rolando S. Tinio (altd.)

German

1 Das Mahl, uns bereitet: ein Zeichen deiner Gnade,
unsern Dank wir vor dich bringen, ein Segenslied dir singen,
führ und leite du uns Gott, dein Licht erhelle diese Tage,
lass uns schmecken dein Güte, lass Neues uns noch wagen.

2 Das Mahl, uns bereitet: ein Zeichen deiner Treue,
wer sich zu dir völlig wendet, im Hunger niemals endet,
führ und leite du uns Gott, dein Güte weite uns're Herzen,
lass uns teilen die Gaben, mit and'ren uns laben.
tr. Dietrich Werner

79

Salwalqulubi

from El Kasilik, Lebanon

♩ = 56

INSTRUMENT (ad lib.)

Sal - wal - qu - lu - bi___ qal - bul - 'ath - ra - 'i
You can con - sole hearts hea - vy with sad - ness.
Du kannst die trau - ri - gen Her - zen er - qui - cken.
Viens con - so - ler la tris - tes - se de nos cœurs.
Con - sue - la los co - ra - zo - nes do - li - dos.

nu - ru - du - ru - bi___ nah - wa - sa - ma - 'i.
You light the path - way___ lead - ing to glad - ness.
Du kannst den Weg zur___ Freu - de uns wei - sen.
Viens é - clai - rer no - tre che - min vers ton bon-heur.
Da - les tu luz, llé - va - los a la vi - da.

English tr. John Campbell; German tr. Päivi Jussila; French tr. Marc Chambron; Spanish tr. Martin Junge

80

Santo es nuestro Dios

Words and music by
Guillermo Cuéllar, El Salvador

San - to, san - to, san - to, san - to, san - to, san - to es nues - tro Dios,
Ho - ly, ho - ly, ho - ly, ho - ly, ho - ly, ho - ly is our God,
Hei - lig, hei - lig, hei - lig, hei - lig, hei - lig, hei - lig un - ser Gott!
Trois fois saint est no - tre Pè - re, trois fois saint est no - tre Dieu.

Se - ñor de to - da la tie - rra, san - to, san - to es nues - tro Dios.
God of all the earth and hea - ven, ho - ly, ho - ly is our God.
Herr des Him - mels und der Er - de. Hei - lig, hei - lig un - ser Gott!
Sei - gneur de tou - te la ter - re; trois fois saint est no - tre Dieu.

San - to, san - to, san - to, san - to, san - to, san - to es nues - tro Dios,
Ho - ly, ho - ly, ho - ly, ho - ly, ho - ly, ho - ly is our God,
Hei - lig, hei - lig, hei - lig, hei - lig, hei - lig, hei - lig un - ser Gott!
Trois fois saint est no - tre Pè - re, trois fois saint est no - tre Dieu.

Se - ñor de to - da la his - to - ria, san - to, san - to es nues - tro Dios.
God of all the earth and hea - ven, ho - ly, ho - ly is our God.
Du bist Herr uns' - rer Ge - schich - te, hei - lig, hei - lig un - ser Gott!
Sei - gneur de tou - te l'hi - stoi - re; trois fois saint est no - tre Dieu.

Que a - com - pa - ña a nues - tro pue - blo, que vi - ve en nues - tras lu - chas, del
Who ac - com - pan - ies our peo - ple, who lives with - in our strug - gles, of
Er be - glei - tet uns' - re Völ - ker und lebt in uns - ren Kämp - fen, auf
Il ac - com - pa - gne son peu - ple dans ses joies et ses pei - nes, et

English tr. Linda McCrae; German tr. Dieter Trautwein; French tr. Marc Chambron

Instruments: guitars, percussion

81 Santo, santo, santo

Unknown, Argentina

Canción ♩ = 64

C G7 Am F G7 F/C C

San - to, san - to, san - to. ¡Mi co - ra - zón te a - do - ra! Mi
Ho - ly, ho - ly, ho - ly. With joy my heart a - dores you! My
Hei - lig, hei - lig, hei - lig. An - be - tend stehn wir vor dir. Aus
Dieu saint, Dieu saint, Dieu saint! Mon cœur, mon cœur t'a - do - re. Mon

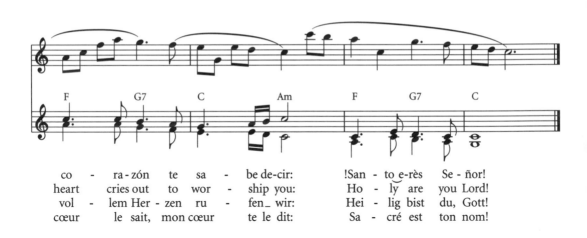

F G7 C Am F G7 C

co - ra - zón te sa - be de - cir: ¡San - to e-rès Se - ñor!
heart cries out to wor - ship you: Ho - ly are you Lord!
vol - lem Her - zen ru - fen_ wir: Hei - lig bist du, Gott!
cœur le sait, mon cœur te le dit: Sa - cré est ton nom!

English tr. Maggie Hamilton; German tr. Dietrich Werner; French tr. Joëlle Gouël

82

Siyabonga

Trad., Southern/East Africa

English

Thank you, Jesus, amen.
Thank you, Jesus, amen.
Thank you, Jesus, amen.
Alleluia, amen.

French

Merci, Jésus, amen.
Merci, Jésus, amen.
Merci, Jésus, amen.
Alléluia, amen.

German

Dank' sei dir, o Jesus.
Dank' sei dir, o Jesus.
Dank' sei dir, o Jesus.
Halleluja. Amen.

Spanish

Gracias, Jesús, amén.
Gracias, Jesús, amén.
Gracias, Jesús, amén.
Aleluya, amén.

83

Seigneur, rassemble-nous

Words and melody by
Dominique Ombrie, France

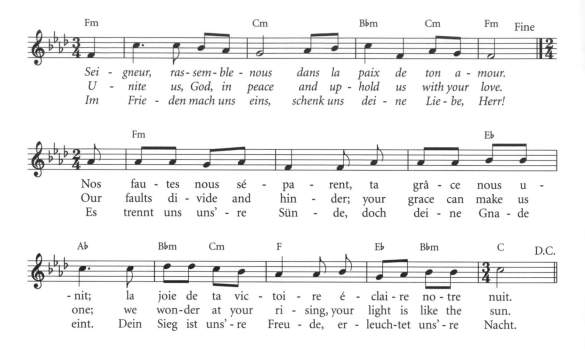

Sei - gneur, ras - sem - ble - nous dans la paix de ton a - mour.
U - nite us, God, in peace and up - hold us with your love.
Im Frie - den mach uns eins, schenk uns dei - ne Lie - be, Herr!

Nos fau - tes nous sé - pa - rent, ta grâ - ce nous u -
Our faults di - vide and hin - der; your grace can make us
Es trennt uns uns' - re Sün - de, doch dei - ne Gna - de

- nit; la joie de ta vic - toi - re é - clai - re no - tre nuit.
one; we won - der at your ri - sing, your light is like the sun.
eint. Dein Sieg ist uns' - re Freu - de, er - leuch - tet uns' - re Nacht.

French

Seigneur, rassemble-nous
dans la paix de ton amour.

1 Nos fautes nous séparent,
ta grâce nous unit;
la joie de ta victoire
éclaire notre nuit.

2 Tu es notre espérance
parmi nos divisions;
plus haut que nos offenses
s'élève ton pardon.

3 Seigneur, vois la misère
des hommes affamés.
Partage à tous nos frères
le pain de l'unité.

4 Heureux le cœur des pauvres
qui cherchent l'unité!
Heureux dans ton royaume
les frères retrouvés.

5 Fais croître en notre attente
l'amour de ta maison;
l'Esprit dans le silence
fait notre communion.

6 Ta croix est la lumière
qui nous a rassemblés;
O joie de notre terre,
tu nous a rachetés.

7 La mort est engloutie,
nous sommes délivrés:
qu'éclate en nous ta vie,
Seigneur ressuscité!

English

Unite us, God, in peace
and uphold us with your love.

1 Our faults divide and hinder;
  your grace can make us one;
  we wonder at your rising,
  your light is like the sun.

2 You are our expectation
  in loneliness and pain;
  your healing and your pardon
  are greater than our sin.

3 Lord, look upon the starving
  and set the captive free.
  Share out among your children
  the bread of unity.

4 How happy are the people
  who strive to be at one,
  who learn to love each other,
  who lay their hatred down.

5 O God, whose silent spirit
  enlightens and endows,
  make us in faith receptive,
  and help us love your house.

6 Your cross will draw together
  the round of humankind;
  in you shall all the people
  their true communion find.

7 Death can no longer hurt us,
  triumphant is your word.
  Let life now grow and blossom,
  O Jesus, risen Lord!

  tr. Fred Kaan

German

Im Frieden mach uns eins,
schenk uns deine Liebe, Herr!

1 Es trennt uns uns're Sünde,
  doch deine Gnade eint.
  Dein Sieg ist uns're Freude,
  erleuchtet unsre Nacht.

2 Du, Herr, bist uns're Hoffnung
  in der Zerrissenheit.
  Wir haben dich beleidigt,
  trotzdem verzeihst du uns.

3 Sieh, wie die Menschen leiden,
  am Hunger, der sie quält!
  Teil aus unter uns allen
  das Brot der Einigkeit!

4 Preist glücklich alle Armen,
  die auf der Suche sind!
  Preist glücklich alle Menschen,
  die zu dir heimgekehrt!

5 Lass in uns allen wachsen
  die Liebe für dein Reich!
  Dein Geist wirkt in der Stille
  Gemeinschaft unter uns.

6 Dein Kreuz wirft helle Strahlen,
  die haben uns vereint.
  O Glück, weil du die Erde
  uns wieder lieben lehrst!

7 Der Tod ist jetzt verschlungen,
  wir Menschen sind befreit!
  Durchdringe unser Leben,
  du auferstand'ner Herr!

  tr. Marlies Flesch-Thebesius

84

Señor, ten piedad de nosotros

Words and music by
Clara Ajo and Pedro Triana, Cuba

Se - ñor,_____ ten pie-dad de no - so - - tros._____
Sen - hor,_____ tem pie-da-de de nós._____

_____ Cris - to,_____ ten pie-dad de no - so - tros._____
_____ Cris - to,_____ tem pie-da-de de nós._____

_____ Se - ñor,_____ ten pie-dad de no - so - tros._____
_____ Sen - hor,_____ tem pie-da-de de nós._____

Sí,_____ ten pie-dad de no - so - tros._____
Sim,_____ tem pie - da - de de nós._____

(Lord, have mercy. Christ, have mercy.)

(Herr, erbarme dich. Jesus Christus, erbarme dich.)

(Seigneur, aie pitié. Jésus-Christ, aie pitié.)

Cumbia

etc.

Percussion: sticks

85

Shang di de gao yang

Mabel Wu, China

qiu lian min_ wo men. Shang di de gao_ yang,
Have_ mer - cy on us. O_ Lamb of God,
er - bar - me dich. O_ Got - tes Lamm,
ô prends pi - tié de nous. O toi, l'A-gneau de Dieu,

ni shi chu qu_ shi ren zui nie de, qiu ci wo men ping an.
you take a - way_ the sin of the world. Grant_ us_ your peace.
du nimmst hin - weg_ die Sün-de der Welt; gib uns Frie - den.
toi qui en - lè - ves le pé-ché du monde, viens nous don-ner ta paix.

German tr. Päivi Jussila; French tr. Marc Chambron

86 | Silence my soul

Rabindranath Tagore, India

Francisco F. Feliciano, Philippines

Ostinato (words repeated in free rhythm)

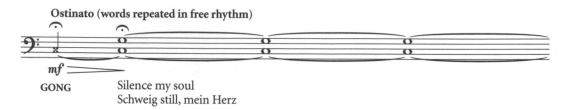

mf

GONG

Silence my soul
Schweig still, mein Herz

Very slow, meditative, in free rhythm

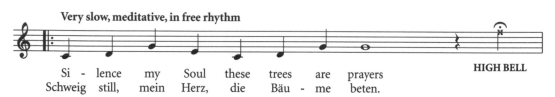

Si - lence my Soul these trees are prayers
Schweig still, mein Herz, die Bäu - me beten.

HIGH BELL

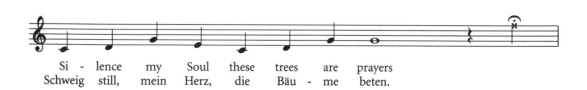

Si - lence my Soul these trees are prayers
Schweig still, mein Herz, die Bäu - me beten.

I asked the Tree
Ich sprach zum Baum.

I asked the Tree
Ich sprach zum Baum:

Tell me__ a-bout God
Er - zähl__ mir von Gott!

Tell me__ a-bout God
Er - zähl__ mir von Gott!

Then it blos - - somed
Und er blüh - - te.

GONG

Ostinato – sung to the words 'Silence my soul', with each singer in free rhythm. Middle C may be added to the ostinato. The ostinato continues throughout until the final gong.

Within the repeats there is freedom to improvise rhythmically, and the voices may enter at different times, overlapping.

The first repeat should be made at least three times. The last phrase may be reserved for the conclusion. The repeat of the final phrase is optional.

Additional verses by Seong-Won Park
Based on the Creation story in Genesis, using the Chinese symbol of days

1 Silence my soul
   the sun is prayer
   I asked the sun
   Tell me about God
   Then it shined

2 Silence my soul
   the moon is prayer
   I asked the moon
   Tell me about God
   Then it glowed

3 Silence my soul
   the fire is prayer
   I asked the fire
   Tell me about God
   Then it surged

4 Silence my soul
   the water is prayer
   I asked the water
   Tell me about God
   Then it rippled

5 Silence my soul
   the tree is prayer
   I asked the tree
   Tell me about God
   Then it blossomed

6 Silence my soul
   the tool is prayer
   I asked the tool
   Tell me about God
   Then it toiled

7 Silence my soul
   the earth is prayer
   I asked the earth
   Tell me about God
   Then it gave life

87

Strengthen for service, Lord

Syriac Liturgy of Malabar
tr. Charles Humphries and Percy Dearmer

JESU DULCIS MEMORIA
Plainchant

Streng - then for ser - vice, Lord, the hands that ho - ly things have
Stär - ke im Dienst, Gott, die Hän - de, die zum Zeug - nis be -
Nos mains, Sei - gneur, ont pris ton pain qu'elles servent aus - si no -
Da fuer - za a las ma - nos, Se - ñor, pues lo sa - gra - do han

ta - ken; let ears that now have heard thy songs_____
- ru - fen. Und lass die Oh - ren, die hö - ren dein Wort,
- tre pro - chain; Nos o - reil - les t'ont en - ten - du,_____
te - ni - do; a - bre los oí - dos a to - do cla - mor,

to cla - mour ne - ver wa - ken. A - - men._____
nicht Lü - ge und Falsch - heit sich öffnen. A - - men._____
qu'el - les t'é - cou - tent tou - jours plus. A - - men._____
pues han oí - do tus can - tos. A - - mén._____

English

1 Strengthen for service, Lord, the hands
that holy things have taken;
let ears that now have heard thy songs
to clamour never waken.

2 Lord, may the tongues which 'Holy' sang
keep free from all deceiving;
the eyes which saw thy love be bright,
thy blessed hope perceiving.

3 The feet that tread thy holy courts
from light do thou not banish;
the bodies by thy Body fed
with thy new life replenish. Amen.

German

1 Stärke im Dienst, Gott, die Hände,
die zum Zeugnis berufen.
Und lass die Ohren, die hören dein Wort,
nicht Lüge und Falschheit sich öffnen.

2 Die Zungen, die singen dir Lob,
befreie von Täuschung und Trug.
Die Augen, die sehn deine Liebe,
erleuchte in ihrer Sicht.

3 Die Füsse, die nachfolgen dir,
mögen nicht gehn in die Irre,
sondern die Gläubigen hier leiten,
dass sie weiterziehen in Pracht. Amen.

tr. Dietrich Werner

French

1 Nos mains, Seigneur, ont pris ton pain
qu'elles servent aussi notre prochain;
Nos oreilles t'ont entendu,
qu'elles t'écoutent toujours plus.

2 Nos langues chantent ton saint nom
qu'elles disent aussi ce qui est bon;
Et nos yeux ont vu ton amour
qu'ils le contemplent pour toujours.

3 Et que nos pieds, qui t'ont suivi,
ne se perdent pas dans la nuit;
Qu'ils conduisent tous les croyants
sur ton chemin resplendissant. Amen.

tr. Marc Chambron

Spanish

1 Da fuerza a las manos, Señor,
pues lo sagrado han tenido;
abre los oídos a todo clamor,
pues han oído tus cantos.

2 Luego de cantarte, Santo,
frena las lenguas de engaño
ojos que han visto tu gloria brillar,
vean la esperanza bendita.

3 Los pies que nos llevan acá,
no tropiecen en la noche;
sigan tus sendas por nuestro mundo,
llévennos a seguirte a tí. Amén.

tr. Martin Junge

88

Slava voveki Tsaryu

Trad., Russian

Trad., Russian melody
arr. Ludmila Garbuzova, Russia

♩ = 68

Sla - va vo - ve - ki Tsa - ryu nye -
Glo - ry to the God of cre - a - tion.
Eh - re sei dir Schöp - fer al - len Le - bens.

- byess - no - mu! Sla - va, sla - - - va!
Glo - ri - a, glo - ria, glo - ri - a!
Eh - re sei Gott, dem e - - wi - gen Gott.

Russian

1 Slava voveki Tsaryu nyebyessnomu!
   Slava, slava!

2 Slava Eesusu Christu I Spaseetyelyu!
   Slava, slava!

3 Slava voveki Sviatomu Dukhu!
   Slava, slava!

4 Slava voveki Triyedyinomu Bogu!
   Slava, slava!

German

1 Ehre sei dir Schöpfer allen Lebens.
   Ehre sei Gott, dem ewigen Gott.

2 Ehre sei dir Christus, Herr und Bruder,
   Retter und Heiland für alle Welt.

English

1 Glory to the God of creation.
   Gloria, gloria, gloria!

2 Glory to the Lord Jesus Christ
   and Saviour for ever. Praise the Lord!

3 Glory to the Spirit, Holy Spirit,
   now and for ever, for ever.

4 Glory to the Trinity, praise the Trinity,
   praise the Triune God!
   tr. Julia Garbuzova

3 Ehre sei dir, Gott, du Geist des Lebens,
   Heiliger Geist, der Leben schafft.

4 Ehre dem dreieinigen, lebend'gen Gott.
   Ehre sei Gott, dem ewigen Gott.
   tr. Dietrich Werner

89 | Te ofrecemos nuestros dones

Edwin Mora, Costa Rica
based on 1 Peter 4: 10,11

Walter Vivares, Argentina

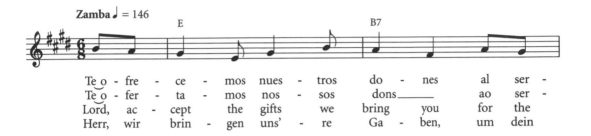

Te o-fre - ce - mos nues - tros do - nes al ser -
Te o-fer - ta - mos nos - sos dons_____ ao ser -
Lord, ac - cept the gifts we bring you for the
Herr, wir brin - gen uns' - re Ga - ben, um dein

-vi - cio de tu rei - no. Te o-fre - ce - mos nues - tra
-vi - ço do teu Rei - no; te o-fer - ta - mos nos - sa
ser - vice of your king - dom; use our lives, our faith, our
Reich hier auf - zu-bau - en. Herr, wir brin - gen un - ser

vi - da por tu cau - sa y por tu a - mor.
vi - da por tua cau - sa e teu a - mor.
ta - lents to make real your peace, your love.
Le - ben, weil du selbst die Lie - be bist.

Portuguese tr. Simei Monteiro; English tr. Fred Kaan; German tr. Dorothea Wulfhorst

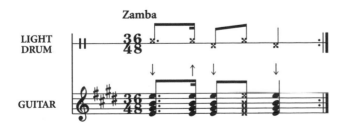

Tembea na Yesu

Trad. Swahili, Kenya

(Walk with Jesus.)

(Geh mit Jesus!)

(Marchons avec Jésus!)

(Camina con Jesús.)

91 | Thuma mina

Trad. Zulu, South Africa

(1. Send me, Lord. 2. I agree, Lord. 3. Thank you, Lord.)

(1. Sende mich, Herr. 2. Ich bin bereit, Herr. 3. Danke, Herr.)

(1. Prends-moi, Seigneur. 2. J'accepte, Seigneur. 3. Merci, Seigneur.)

(1. Envíame, Señor. 2. Estoy listo/a, Señor. 3. Gracias, Señor.)

The deer longs for pure flowing streams

George Mulrain, Trinidad
based on Psalm 42

Trad., Caribbean

The deer longs for pure flow-ing streams;
my thirst-y soul longs for the Lord:
I have been weep-ing all day and all night,
for they say, 'Tell us, when will your God keep his word?'

Wie der Hirsch lechzt nach fri - schem Wasser
so schreit mei-ne See-le zu dir.
Mei - ne Trä-nen sind Spei-se für mich Tag und Nacht,
denn sie sa - gen: Wo ist nun dein Gott?

English

The deer longs for pure flowing streams;
my thirsty soul longs for the Lord:
I have been weeping all day and all night, for they say,
'Tell us, when will your God keep his word?'

So why are you downcast, my soul;
and why are you crying alone?
You must remember to put all your trust in your God,
and to praise him for what he has done.

By day the Lord shows me his love,
his blessings surround me at night:
so I will tell out his faithfulness, sing of his grace,
and, rejoicing, I'll walk in his light!

German

Wie der Hirsch lechzt nach frischem Wasser
so schreit meine Seele zu dir.
Meine Tränen sind Speise für mich Tag und Nacht,
denn sie sagen: Wo ist nun dein Gott?

Was zagest du so, meine Seele
und bist so unruhig in mir.
Harre auf Gott, denn du wirst ihm noch danken,
dass er Hilfe ist, Rat und mein Gott.

Am Tage zeigt Gott seine Güte,
des Nachts sing und bet ich zu ihm.
So will ich singen und danken dem lebend'gen Gott,
dass er Hilfe ist, Rat und mein Gott.
tr. Dietrich Werner

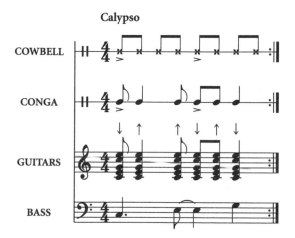

The Great Thanksgiving
Holy, holy, holy Lord

LAND OF REST
Trad., USA

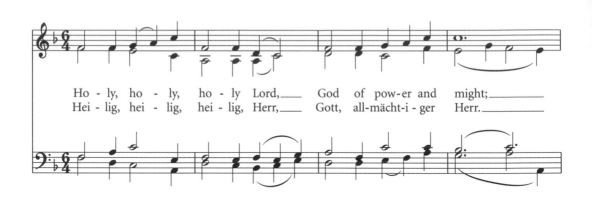

Ho - ly, ho - ly, ho - ly Lord,___ God of pow-er and might;___
Hei - lig, hei - lig, hei - lig, Herr,___ Gott, all-mächt-i - ger Herr.___

heaven and earth_ are full of your glo - ry. Ho - san - na in__ the high - est.
Al - le Land_ sind voll dei-ner Eh - re. Ho - sian - na in__ der Hö - he.

Bless - ed__ is he who comes in the name of the Lord.___ Ho -
Hochge - lobt_ sei, der da kommt im Na - men__ des Herrn.___ Ho -

-san - na in___ the high - est,___ ho - san - na in___ the high - est.
-sian - na,___ ho - sian - na,___ ho - sian - na in___ der Hö - he.

German tr. Päivi Jussila

(Saint, saint, saint est l'Éternel, Dieu de l'univers! Le ciel et la terre sont remplis de ta gloire. Hosanna au plus haut des cieux. Béni soit celui qui vient au nom du Seigneur. Hosanna au plus haut des cieux.)

(Santo, santo, santo es el Señor, Dios del universo. Llenos están el cielo y la tierra de tu gloria. Hosanna en el cielo. Bendito el que viene en el nombre del Señor. Hosanna en el cielo.)

93b
Christ has died

Christ has died.___ Christ is ris - en.__ Christ will come a - gain._____
Chris - tus starb,___ ist er-stan-den,_ kommt in Herr - lich - keit._____

Christ has died.___ Christ is ris - en.__ Christ will come a - gain.___
Chris - tus starb,___ ist er-stan-den,_ kommt in Herr - lich - keit.___

German tr. Angelika Joachim

(Christ est mort. Christ est ressuscité. Christ reviendra.)

(Cristo ha muerto. Cristo ha resucitado. Cristo vendrá otra vez.)

93c
Amen

A - men,___ a - men,_ a - - - men.___

94

Tú eres amor

Words and music by
Teresita Savall, Argentina

Canon

Baguala ♩ = 170

Tú   e - res a - mor,   tú   e - res bon - dad,   Se -
You,   Source of all love,   you,   Giv - er of good,   O
Die   Lie - be ist dein,   die   Gü - te ist dein.   Um
Toi,   qui es l'a - mour,   toi,   qui es bon - té.   Sei -

- ñor,   da - nos la paz,   que el mun - do no pue - de dar.
God,   grant us your peace,   the peace that the world can't give.
Frie -   den bit - ten wir: Der Frie - den kommt stets von dir.
- gneur,   don - ne la paix   que seul tu peux nous don - ner.

English tr. Fred Kaan; German tr. Dorothea Wulfhorst; French tr. Marc Chambron

GUITAR

95

Those who wait upon the Lord

based on Isaiah 40: 31

Trad., North America

Those who wait u - pon the Lord shall re - new their strength,
Die auf ih - ren Herrn ver - traun, schöp-fen neu - e Kraft.

they shall mount up on wings as ea - gles.____
Er ver - leiht Flü - gel wie den Ad - lern,____

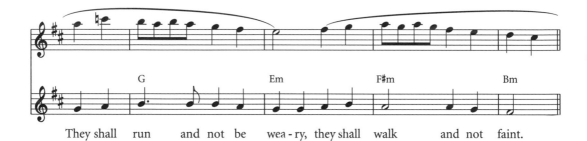

They shall run and not be wea - ry, they shall walk and not faint.
dass sie lau - fen, nicht er - mü-den, dass sie gehn vol - ler Kraft.

Help us, Lord, help us, Lord, in your way.____
Hilf uns, Herr, hilf uns, Herr, auf dem Weg.____

English

1 Those who wait upon the Lord
   shall renew their strength,
   they shall mount up on wings as eagles.
   They shall run and not be weary,
   they shall walk and not faint.
   Help us, Lord, help us, Lord, in your way.

2 Those who serve a suff'ring world …

3 Those who sow the seeds of hope …

4 Those who work for peace on earth …

German

1 Die auf ihren Herrn vertraun
   schöpfen neue Kraft.
   Er verleiht Flügel wie den Adlern,
   dass sie laufen, nicht ermüden,
   dass sie gehn voller Kraft.
   Hilf uns, Herr, hilf uns, Herr, auf dem Weg.

2 Die bereit zum Dienen sind …

3 Die die Saat der Hoffnung sä'n …

4 Die die Saat des Friedens sä'n …
                              tr. Angelika Joachim

(Ésaie 40: 31 Ceux qui espèrent dans le Seigneur retrempent leur énergie: ils prennent de l'envergure comme des aigles, ils s'élancent et ne se fatiguent pas, ils avancent et ne faiblissent pas!)

(Isaías 40: 31. Pero los que confían en el Señor tendrán siempre nuevas fuerzas y podrán volar como las águilas; podrán correr sin cansarse y caminar sin fatigarse.)

96

Truth is our call

Words and music by
Rima Nasir Tarazi, Palestine

Truth is our call and jus - tice our claim. The will of our God is our
Wahr - heit und Recht, da - zu du uns rufst. Des Le - bens— Ziel nur dein

van - guard and aim; the God of us all, of mer - cy and love, of
Wil - le be - stimmt. Du Gott al - ler Menschen, voll Lie - be und Gnad, der

free - dom and peace for all of hu - man - kind.
Frei - heit und Frie - den will für je - des Kind.

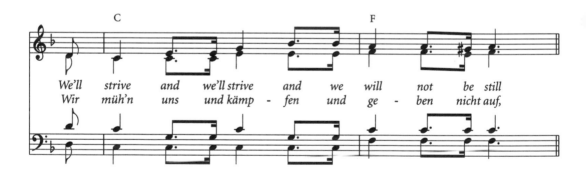

We'll strive and we'll strive and we will not be still
Wir müh'n uns und kämp - fen und ge - ben nicht auf,

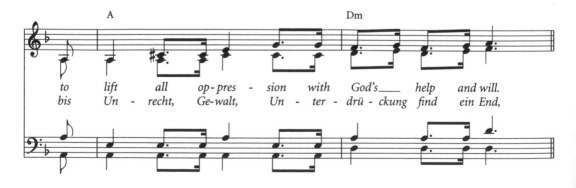

to lift all op - pres - sion with God's— help and will.
bis Un - recht, Ge-walt, Un - ter - drü - ckung find ein End,

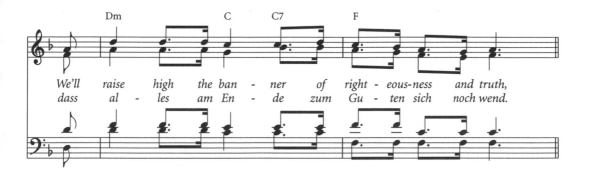

We'll raise high the ban - ner of right - eous-ness and truth,
dass al - les am En - de zum Gu - ten sich noch wend.

we'll strive and we'll strive and we will not be still.
Wir müh'n uns, zu än - dern was am Wel - ten Lauf.

English

1 Truth is our call and justice our claim.
The will of our God is our vanguard and aim;
the God of us all, of mercy and love,
of freedom and peace for all of humankind.
We'll strive and we'll strive and we will not be still
to lift all oppression with God's help and will.
We'll raise high the banner of righteousness and truth,
we'll strive and we'll strive and we will not be still.

2 We shall not give in to fear or to hate;
we will speak the truth and we'll strive to be just.
With love we will stir the conscience of the world;
with patience and faith we'll save our home and land.

German

1 Wahrheit und Recht, dazu du uns rufst.
Des Lebens Ziel nur dein Wille bestimmt.
Du Gott aller Menschen, voll Liebe und Gnad,
der Freiheit und Frieden will für jedes Kind.
Wir müh'n uns und kämpfen und geben nicht auf,
bis Unrecht, Gewalt, Unterdrückung find ein End,
dass alles am Ende zum Guten sich noch wend.
Wir müh'n uns, zu ändern was am Welten Lauf.

2 Frei macht uns Gott von Angst oder Hass.
Sein Recht, seine Wahrheit ist unser Mass.
Mit Liebe verwandeln wir diese Welt.
Geduld und Hoffnung unser Land erhält.
tr. Dietrich Werner

97

Tua Palavra na vida

Words and music by
Simei Monteiro, Brazil

Tu - a Pa - la - vra na vi - - - da
Es tu Pa - la - bra en la vi - - - da
Your Word in our lives, e - ter - - - nal,
Dein Wort er - füllt sich im Le - - - ben:

to intro.

(*verses overleaf*)

173

Portuguese

1 Tua Palavra na vida
é fonte que jamais seca,
água que anima e restaura
todos que a queiram beber.

2 Tua Palavra na vida
é qual semente que brota;
torna-se bom alimento,
pão que não há de faltar.

3 Tua Palavra na vida
é espelho que bem reflete,
onde nos vemos, sinceros,
como a imagem de Deus.

4 Tua Palavra na vida
é espada tão penetrante
que revelando as verdades
vai renovando o viver.

5 Tua Palavra na vida
é luz que os passos clareia,
para que ao fim no horizonte
se veja o Reino de Deus.

Spanish

1 Es tu Palabra en la vida
fuente que nunca se seca,
agua que anima y restaura
a quienes quieran beber.

2 Es tu Palabra en la vida
como semilla que brota;
llega a ser buen alimento,
pan que no nos faltará.

3 Es tu Palabra en la vida
espejo que bien refleja,
donde nos vemos, sinceros,
como la imagen de Dios.

4 Es tu Palabra en la vida
espada tan penetrante
que revelando verdades
va renovando el vivir.

5 Es tu Palabra en la vida
luz que los pasos aclara
y al fin, en el horizonte,
nos muestra el Reino de Dios.
tr. Jorge Rodríguez
and Pablo Sosa

English

1 Your Word in our lives, eternal,
  it is a clear fountain flowing;
  water that gives strength and courage
  to all who draw near and drink.

2 Your word in our lives, eternal,
  seed of the Kingdom that's growing;
  it becomes bread for our tables,
  food for the feast without end.

3 Your word in our lives, eternal,
  becomes the mirror where we see
  the true reflection of ourselves:
  children and image of God.

4 Your word in our lives, eternal,
  it is a sharp two-edged sword;
  dividing our lies from your truth,
  it's bringing new life to all.

5 Your word in our lives, eternal,
  is light that shines on the long road
  that leads us to the horizon
  and the bright Kingdom of God.
  tr. Sonya Ingwersen

German

1 Dein Wort erfüllt sich im Leben:
  Quelle, die niemals vertrocknet,
  Wasser, die Kräfte zu wecken
  jedem, der nahkommt und trinkt.

2 Dein Wort erfüllt sich im Leben:
  Samen, der keimt, sich entfaltet.
  Nahrung, die froh wir empfangen,
  Brot, das uns niemals mehr fehlt.

3 Dein Wort erfüllt sich im Leben:
  Spiegel, darin uns zu schauen,
  aufrichtig jetzt zu begreifen,
  alle schuf Gott als sein Bild.

4 Dein Wort erfüllt sich im Leben:
  Schwert, das tief eindringt und scharf ist,
  Wahrheit, die weh tut, ans Licht bringt,
  Leben erneuert und heilt.

5 Dein Wort erfüllt sich im Leben:
  Licht, das die Schritte uns hell macht,
  lässt uns, was kommt, klar erkennen,
  ewig wird währen dein Reich.
  tr. Dieter Trautwein

98

Utukufu

Trad. kiSwahili, Kenya

♩ = 112

LEADER / ALL

Utu-ku-fu | Utu-ku-fu kwa mu-ngu juu
Sing glo-ry! | Glo-ry be to God on high.
Singt: Eh-re, | Eh-re sei dir in der Höh.

LEADER / ALL

A-ma-ni | A-ma-ni du ni-a-ni
Let peace reign! | Peace be to all cre-a-tion.
Singt: Frie-den, | Frie-den gibst du uns al-len.

LEADER / ALL

K'we-nye ma-pe-nzi | K'we-nye ma-pe-nzi me-ma
Let love in-spire us! | Let love be gen-'rous and full.
Du bist die Lie-be, | e-wi-ge Lie-be bist du.

LEADER / ALL

K'we-nye ma-pe-nzi | K'we-nye ma-pe-nzi me-ma.
Let love in-spire us! | Let love be gen-'rous and full.
Du bist die Lie-be, | e-wi-ge Lie-be bist du.

English tr. Fred Kaan; German tr. Angelika Joachim and Päivi Jussila

99

Våga vara den du i Kristus är

Anders Frostenson

Roland Forsberg, Sweden

Vå - ga va - ra den du i Kris - tus är,
Take the risk, be all that you are in Christ:
Wag's und sei doch, was du in Chri - stus bist,

den i hans tan - ke, den i hans kär - lek,
Christ in your think - ing, Christ in your lov - ing,
in sei - nem Ur - teil, in sei - ner Lie - be,

den i hans ö - gas e - vi - ga ljus du är.
let the tran - scen - dent light of your eyes be Christ's.
in sei - nes Au - ges e - wi - gem Licht schon bist!

Swedish

1 Våga vara den du i Kristus är,
   den i hans tanke,
   den i hans kärlek,
   den i hans ögas eviga ljus du är.

2 Intet äger du som ej han dig ger.
   Kravet och kraften,
   viljan och verket
   strömmar ur samma källa från
        bergen ner.

3 Skuld och rädsla trycker dig inte mer.
   Nu är du fri att
   älska och tjäna
   dem som du möter, Jesus som
        bor i dem.

4 Redan är du den du en gång skall bli:
   dömd och benådad,
   död och uppstånden,
   älskad och ett med honom som
        gjort dig fri.

English

1 Take the risk, be all that you are in Christ:
   Christ in your thinking,
   Christ in your loving,
   let the transcendent light of your eyes
        be Christ's.

2 Christ has given all that you have today:
   calling and courage,
   brave will and actions
   flow from the mountain-spring of the
        Christ always.

3 Shed the burden of all your guilt and fear;
   now you are free
   to love and to serve
   the stranger and neighbour: in them the
        Christ is here.

4 You are now all that you will onc day be:
   judged and forgiven,
   dead and yet risen,
   cherished, and one with Christ who has
        set you free.
        tr. Maggie Hamilton

German

1 Wag's und sei doch, was du in Christus bist,
   in seinem Urteil,
   in seiner Liebe,
   in seines Auges ewigem Licht schon bist!

2 Gar nichts hast du, was er nicht selbst dir gab.
   Anspruch und Antwort,
   Wollen und Wirken
   strömen aus gleicher Quelle den Berg hinab.

3 Schuld und Ängste lasten nicht mehr auf dir.
   Nun bist du frei,
   zu dienen und lieben,
   wen du auch triffst und Jesus in ihm, in ihr.

4 Jetzt schon bist du, der einmal werden wird:
   schuldig und heilig,
   tot und erstanden,
   freigeliebt, eins mit Ihm, der dich heimgeführt.
        tr. Jürgen Henkys

100

Voice-over God, our Lover and Creator

Fred Kaan, Netherlands/England

Anfinn Øien, Norway

Voice-ov-er God, our Lov-er and Cre-a-tor,
Gott, der das Le-ben schuf, uns A-tem schenk-te,

whose Word trans-formed dis-or-der in-to peace,
der die-se Welt aus Chaos zum Frie-den rief.

who gave to all that is: a name, a call-ing,
Du kennst mit Na-men je-des Le-be-we-sen.

by whom all hu - man wis - dom was re - leased,
Voll dei - ner Weis - heit ist, was lebt und wirkt.

God! give us con - fi - dence in liv - ing, lov - ing,
Gott, gib uns Mut zu le - ben im Ver - trau - en,

the mind, the mood to make this life a feast.
das Le - ben sei ein Spie - gel dei - ner Gnad.

(verses overleaf)

English

1 Voice-over God, our Lover and Creator,
   whose Word transformed disorder into peace,
   who gave to all that is: a name, a calling,
   by whom all human wisdom was released,
   God! give us confidence in living, loving,
   the mind, the mood to make this life a feast.

2 Com-panion* Christ, in human terms among us,
   let your redeeming, freeing presence teach
   and help us not to measure those around us
   by culture, race, by gender, creed or speech,
   to look for that which is of God within them,
   reach out, give thanks, for what is good in each.

3 Ennobling Spirit, source of strength and passion,
   inspire our hearts to hear the voiceless cry,
   our will to break the shackles of injustice,
   take down all barriers that demean, divide;
   to give the hungry bread, the homeless shelter:
   life in all fullness, that will *never* die.

* Companion: from the Latin *cum* (together with) and *panis* (bread). A companion or a company is a person or persons with whom you share bread!

German

1 Gott, der das Leben schuf, uns Atem schenkte,
   der diese Welt aus Chaos zum Frieden rief.
   Du kennst mit Namen jedes Lebewesen.
   Voll deiner Weisheit ist, was lebt und wirkt.
   Gott, gib uns Mut zu leben im Vertrauen,
   das Leben sei ein Spiegel deiner Gnad.

2 Christ, unser Bruder, Mensch an uns'rer Seite,
   sei gegenwärtig mitten unter uns,
   dass weder Rasse, Religion noch Farbe
   verstell den Blick für Würde, Menschenrecht,
   dass wir dein Bild und deine Spur erkennen
   in jedem Menschen, den du liebst wie mich.

3 Geist, Gottes Kraft und Stärke auch im Leiden,
   mach unsere Herzen weit, wenn Stumme schrei'n,
   mach uns'ren Willen fest, ein Nein zu sagen,
   zu allem Unrecht, allem, was uns trennt,
   Brot sei dem Hungrigen und Schutz gegeben,
   so dass von Lebensfülle alle spür'n.
                    tr. Dietrich Werner

101

Vai com Deus!

Words and music by
Simei Monteiro, Brazil

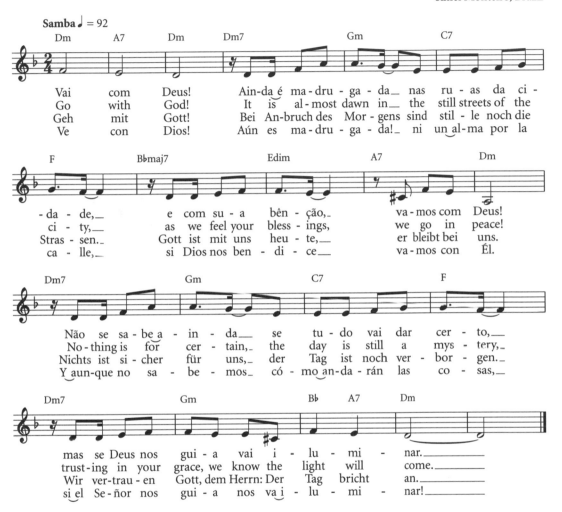

Samba ♩ = 92

Dm / A7 / Dm / Dm7 / Gm / C7

Vai com Deus! Ain-da é ma-dru-ga-da__ nas ru-as da ci-
Go with God! It is al-most dawn in__ the still streets of the
Geh mit Gott! Bei An-bruch des Mor-gens sind stil-le noch die
Ve con Dios! Aún es ma-dru-ga-da! ni un al-ma por la

F / Bbmaj7 / Edim / A7 / Dm

-da-de,__ e com su-a bên-ção,__ va-mos com Deus!
ci-ty,__ as we feel your bless-ings, we go in peace!
Stras-sen.__ Gott ist mit uns heu-te,__ er bleibt bei uns.
ca-lle,__ si Dios nos ben-di-ce__ va-mos con Él.

Dm7 / Gm / C7 / F

Não se sa-be a-in-da__ se tu-do vai dar cer-to,__
No-thing is for cer-tain,__ the day is still a mys-tery,__
Nichts ist si-cher für uns,__ der Tag ist noch ver-bor-gen.__
Y aun-que no sa-be-mos__ có-mo an-da-rán las co-sas,__

Dm7 / Gm / Bb / A7 / Dm

mas se Deus nos gui-a vai i-lu-mi-nar._____
trust-ing in your grace, we know the light will come._____
Wir ver-trau-en Gott, dem Herrn: Der Tag bricht an._____
si el Se-ñor nos gui-a nos va i-lu-mi-nar!_____

German tr. Päivi Jussila; Spanish tr. Pablo Sosa

Samba (in combination)

Instruments: guitars, percussion

102

Vos sos el destazado en la cruz

Words and music by
Carlos Mejia Godoy, Nicaragua

♩. = 78

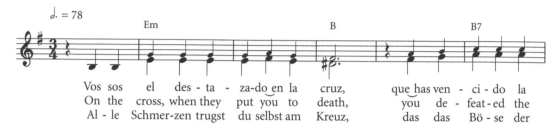

Vos sos el des - ta - za-do en la cruz, que has ven - ci - do la
On the cross, when they put you to death, you de - feat - ed the
Al - le Schmer-zen trugst du selbst am Kreuz, das das Bö - se der

mal-dad del mun - do,___ de - nun - cian-do al in - jus-to o-pres - sor,
for - ces of e - vil,___ you con-demned the op - pres-sors' in - justice,
Welt ü - ber - wun - den.___ Un - ter - drü-ckung hast du bloss-ge - stellt,

le - van - tan - do del pol-vo a los po - bres. Te pe - di - mos
raised the poor from the dust of de - press - ion. Hear our pro - test,
aus dem Staub auf - ge - rich-tet die Ar - men. Wir___ ru - fen,

que nos oi - gas,___ que es-cu - ches el cla - mor de tu pue - blo. Te pe -
heed our an - guish,___ and res-pond to the hopes of your peo - ple. Hear our
drän - gen, bit - ten,___ dass die Kla-gen du hörst dei-nes Vol - kes. Wir___

- di - mos que nos oi - gas,___ que es-cu - ches el cla - mor de tu pue - blo.
pro - test, heed our an - guish,___ and res-pond to the hopes of your peo - ple.
ru - fen, drän - gen, bit - ten,___ dass die Kla-gen du hörst dei-nes Vol - kes.

Spanish

1 Vos sos el destazado en la cruz,
  que has vencido la maldad del mundo,
  denunciando al injusto opresor,
  levantando del polvo a los pobres.
  Te pedimos que nos oigas,
  que escuches el clamor de tu pueblo.
  Te pedimos que nos oigas,
  que escuches el clamor de tu pueblo.

2 Vos sos el destazado en la cruz,
  masacrado por los poderosos;
  hoy derramas tu sangre también
  en la sangre de nuestros caídos.
  Te pedimos …

3 Vos sos el destazado en la cruz,
  que construyes la paz con justicia.
  Ayúdanos a no desmayar,
  a luchar por que venga tu Reino.
  Que tu paz llegue a nosotros
  cuando hagamos brotar la justicia.
  Que tu paz llegue a nosotros
  cuando hagamos brotar la justicia.

English

1 On the cross, when they put you to death,
  you defeated the forces of evil,
  you condemned the oppressors' injustice,
  raised the poor from the dust of depression.
  Hear our protest, heed our anguish,
  and respond to the hopes of your people.
  Hear our protest, heed our anguish,
  and respond to the hopes of your people.

2 On the cross, when they put you to death
  (they—the merchants of ill and destruction),
  you were one with all victims of torture,
  loved ones, shedding their life-blood for others.
  Hear our protest …

3 Yet the cross, where you died, is today
  sign and symbol of life, peace and justice,
  stands among us and calls us to service:
  let your Kingdom of love rule creation!
  Let your peace inspire, unite us
  in pursuit of life's fullness and justice.
  Let your peace inspire, unite us
  in pursuit of life's fullness and justice.
                              tr. Fred Kaan

German

1 Alle Schmerzen trugst du selbst am Kreuz,
  das das Böse der Welt überwunden.
  Unterdrückung hast du blossgestellt,
  aus dem Staub aufgerichtet die Armen.
  Wir rufen, drängen, bitten,
  dass die Klagen du hörst deines Volkes.
  Wir rufen, drängen, bitten,
  dass die Klagen du hörst deines Volkes.

2 Alle Schmerzen trugst du selbst am Kreuz,
  ein Opfer der Mächt'gen geworden.
  Und das Blut, das gekostet dein Tod,
  uns're Märtyrer heute verlieren.
  Wir rufen …

3 Alle Schmerzen trugst du selbst am Kreuz,
  durch das Frieden wächst in Recht und Freiheit.
  Hilf, dass lähmend in uns nicht die Angst,
  sondern stark unsere Hoffnung:
        dein Reich kommt!
  Deinen Frieden gib uns heute,
  dass wir bleiben die Zeugen des Rechtes.
  Deinen Frieden gib uns heute,
  dass wir bleiben die Zeugen des Rechtes.
              tr. Dietrich Werner

103

Wa Emimimo

The Church of the Lord (Aladura), Nigeria

Deliberately ♩ = 80

ALL

Wa wa wa E - mi - mi - mo,
Come, O Ho - ly Spir - it, come,
Komm, o komm, Hei - li - ger Geist,
O viens, Es - prit,___ viens

LEADER

E - mi - o - lo - ye.
Ho - ly Spir - it, come.
Geist der Wahr - heit
Es - prit de sa - gesse

Wa wa wa A - lag - ba - ra,
Come, al - migh - ty Spir - it, come,
komm, o komm, du Geist voll Kraft,
O viens, puis - sant Es - prit, viens,

A - lag - ba - ra - me - ta.
al - might - y Spir - it, come.
du Geist der Ei - nig - keit,
Puis - san - te Tri - ni - té

Wa - o, wa - o, wa - o.___
Come,_ come,_ come.___
komm,_ komm,_ komm.___
viens,_ viens,_ viens.___

E - mi - mi - mo.
O Spir - it, come.
Hei - li - ger Geist.
O Es - prit, viens.

English tr. I-to Loh; German tr. Wolfgang Leyk; French tr. Joëlle Gouël

SHAKER

DRUM 1

DRUM 2
or GONG

104 We believe, Maranatha

based on Revelation 22: 5,20

Francisco F. Feliciano, Philippines

German tr. Dieter Trautwein; French tr. Marc Chambron

105

We shall go out with hope of resurrection

June Boyce-Tillman, England

LONDONDERRY AIR
Trad., Northern Ireland

bold - ly,   tales of a love   that will not let us   go._____ We'll sing our
Hoff-nung, Bot-schaft der Liebe, die nie-mand je-mals aufgibt.____ Wir sing'n das

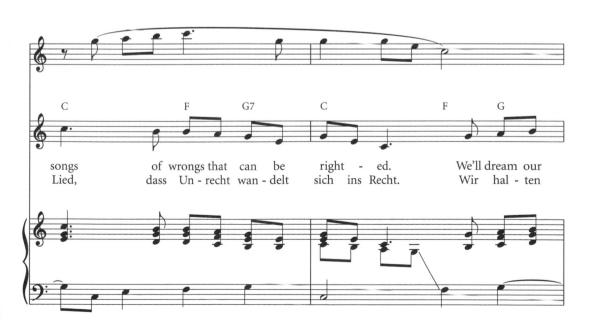

songs   of wrongs that can   be   right - ed.   We'll dream our
Lied,   dass Un-recht wan-delt   sich ins Recht.   Wir hal-ten

dreams   of hurts that can be   healed._____ We'll weave a cloth   of all the world u -

fest   den Traum, dass Wun-den   heilen._____ Wir stri-cken wei - ter an dem Netz der

- ni - ted   with - in the vi - sion of a Christ who sets us   free._____

Hoff - nung,   dass ei - ne Welt in Chris-tus wird, der macht uns   frei._____

English

1 We shall go out with hope of resurrection;
  we shall go out, from strength to strength go on.
  We shall go out and tell our stories boldly,
  tales of a love that will not let us go.
  We'll sing our songs of wrongs that can be righted.
  We'll dream our dreams of hurts that can be healed.
  We'll weave a cloth of all the world united
  within the vision of a Christ who sets us free.

2 We'll give a voice to those who have not spoken.
  We'll find a word for those whose lips are sealed.
  We'll make the tunes for those who sing no longer,
  vibrating love alive in every heart.
  We'll share our joy with those who still are weeping,
  chant hymns of strength for hearts that break in grief.
  We'll leap and dance the resurrection story,
  including all within the circles of our love.

German

1 Wir stehen auf in der Kraft des Auferstand'nen,
  wir ziehen aus gestärkt durch seine Macht.
  Wir stehen auf mit Geschichten seiner Hoffnung,
  Botschaft der Liebe, die niemand jemals aufgibt.
  Wir sing'n das Lied, dass Unrecht wandelt sich ins Recht.
  Wir halten fest den Traum, dass Wunden heilen.
  Wir stricken weiter an dem Netz der Hoffnung,
  dass eine Welt in Christus wird, der macht uns frei.

2 Wir geben Stimme allen, denen das Wort verboten.
  Wir rufen laut für alle, die verstummt sind.
  Wir singen neu für alle, denen das Lied vergangen,
  dass Liebe wächst erneut in jedem Herz.
  Wir teilen Freude mit all denen, die noch weinen.
  Wir singen Lieder zu stärken, die voll Kummer.
  Wir tanzen, feiern das Fest des Auferstand'nen,
  weiten die Kreise unserer Liebe nach aussen hin.
  tr. Dietrich Werner

106

Wij geloven één voor één

Joke Ribbers

Bernard Smilde, Netherlands

Wij ge - lo - ven één voor één en ook sa - men:
We be - lieve, as one by one and to - ge - ther;
Lasst uns glau - ben ganz al - lein, und zu - sam - men,
C'est la foi de tout chré - tien, c'est la nô - tre:

de Heer is God en an - ders geen. A - men, a - men.
the Lord is God, there is none else, a - men, a - men.
dass Gott der Herr ist, nie - mand sonst. A - men. A - men.
Dieu tient le mon - de dans ses mains. A - men, a - men.

Dutch

1 Wij geloven één voor één
en ook samen:
de Heer is God en anders geen.
Amen, amen.

2 Wij geloven in de naam
Jezus Christus,
gestorven en weer opgestaan.
Halleluja!

3 Wij geloven dat de Geest
ook nog heden
de wereld en onszelf geneest.
Vrede, vrede.

English

1 We believe, as one by one
and together;
the Lord is God, there is none else,
amen, amen.

2 We believe in Jesus's name,
Christ among us,
condemned to death, but raised to life:
Alleluia!

3 We believe that even now
God the Spirit
will heal the world and heal us all:
Peace be with us!

tr. Fred Kaan

German

1 Lasst uns glauben ganz allein,
und zusammen,
dass Gott der Herr ist, niemand sonst.
Amen. Amen.

2 Lasst uns glauben Jesus Christ,
unser Heiland,
gestorben und erstanden ist.
Amen. Amen.

3 Lasst uns glauben, dass der Geist,
Geist des Lebens,
die Welt verwandelt und uns heilt.
Amen. Amen.

tr. Dietrich Werner

French

1 C'est la foi de tout chrétien,
c'est la nôtre:
Dieu tient le monde dans ses mains.
Amen, amen.

2 Nous croyons en Jésus-Christ,
mort sur la croix:
Pour toujours il a repris vie.
Alléluia!

3 Nous croyons en l'Esprit Saint,
qui, sans cesse,
vient pour guérir tous les humains,
et le monde.

tr. Marc Chambron

107

Word of justice (Litany of the Word)

Words and music by
Bernadette Farrell, England

Word of jus - tice,   Al - le - lu - ia,
Wort des Rech - tes,   Hal - le - lu - ja,
O Pa - ro - le,   Al - lé - lu - ia,

come to dwell here.   Ma - ra - na - tha!
nimm hier Woh - nung.   Ma - ra - na - tha!
viens par - mi nous,   Ma - ra - na - tha!

English

1 Word of justice, *Alleluia,* come to dwell here. *Maranatha!*
2 Word of mercy, *Alleluia,* live among us. *Maranatha!*
3 Word of power, *Alleluia,* live within us. *Maranatha!*
4 Word of freedom, *Alleluia,* save your people. *Maranatha!*
5 Word of healing, *Alleluia,* heal our sorrow. *Maranatha!*
6 Word of comfort, *Alleluia,* bring us hope now. *Maranatha!*
7 Word of gladness, *Alleluia,* fill our hearts now. *Maranatha!*
8 Word of wisdom, *Alleluia,* come renew us. *Maranatha!*

German

1 Wort des Rechtes, *Halleluja,* nimm hier Wohnung. *Maranatha!*
2 Wort der Gnade, *Halleluja,* sei uns Leben. *Maranatha!*
3 Wort der Stärke, *Halleluja,* sei uns Beistand. *Maranatha!*
4 Wort der Freiheit, *Halleluja,* bring Erlösung. *Maranatha!*
5 Wort der Heilung, *Halleluja,* heile Wunden. *Maranatha!*
6 Wort des Trostes, *Halleluja,* stärke Hoffnung. *Maranatha!*
7 Wort der Freude, *Halleluja,* füll die Herzen. *Maranatha!*
8 Wort der Weisheit, *Halleluja,* komm, erneu're. *Maranatha!*
     tr. Dorothea Wulfhorst

French

1 O Parole, *Alléluia,* viens parmi nous, *Maranatha!*
2 Tu pardonnes, *Alléluia,* tu te donnes, *Maranatha!*
3 Tu es force, *Alléluia,* et puissance, *Maranatha!*
4 Tu libères, *Alléluia,* tu nous sauves, *Maranatha!*
5 Ta guérison, *Alléluia,* nous l'attendons, *Maranatha!*
6 Tu consoles, *Alléluia,* nos tristesses, *Maranatha!*
7 Allégresses, *Alléluia,* et tendresse, *Maranatha!*
8 O sagesse, *Alléluia,* ô promesse, *Maranatha!*
     tr. Marc Chambron

INSTRUMENTAL INTERLUDE

108

Xin Chua thuong xot chung con

Unknown, Vietnam

German tr. Inge Klaas

(Seigneur, aie pitié de nous. Jésus-Christ, aie pitié. Seigneur, aie pitié de nous.)

(Señor, ten piedad. Jesucristo, ten piedad. Señor, ten piedad.)

109 | Ya hamalaLah

Yusuf Khill, Palestine/Israel

Ya_ ha-ma - la - Lah al - ha-mel kha - ta - yal - 'a_____ lam ir - ham____ na.
O_ Lamb of God, you take_ a - way the sins of the world, have mer-cy on us.
Du Got-tes-lamm nimmst hin-weg al - le Schuld auf Er - den, er - bar - me dich.

Ya_ ha - ma - la - Lah al - ha - mel kha - ta - yal - 'a_____ lam im -
O_ Lamb of God, you take___ a - way the sins_ of the world, grant
Du_ Got-tes-lamm nimmst hin - weg al - le Schuld aur Er - den, bring

-nah - na - sa - lam, im - nah - na - sa - lam, im - nah - na - sa - lam.
us your_ peace, grant us your_ peace, grant us____ your peace.
Frie - den_ uns, bring Frie - den_ uns, bring Frie - den zu uns.

German tr. R. Renner and U. Schwendler

110

Yarabba ssalami

Unknown, Palestine

♩ = 78

Ya - ra - bba ssa - la - mi   am - ter a - lay - na ssa - lam,
God of peace, in your wis - dom   give us the will to seek peace;
Du__ Gott des__ Frie - dens,   giess dei - nen Frie - den auf uns,
Toi,__ Dieu de la paix____   en - voie ta paix sur nous tous.
Tu, oh Dios, de la paz____   da - nos a - mor por tu paz;

ya - ra - bba ssa - la - mi   im la' qu - lu - ba - na sa - lam.
God of peace and of heal - ing,   fill with your peace ev - ery heart!
du__ Gott des__ Frie - dens,   fül - le mit Frie-den un - ser Herz.
Toi,__ Dieu de la paix____   com - ble nos cœurs de ta paix.
Tu, oh Dios, de la paz____   en nues-tras vi - das siem-bra paz.

English tr. Fred Kaan; German tr. Renate Schiller; Spanish tr. Martin Junge

Sung unaccompanied, or with an oud or guitar playing the melody in repeated notes.

111

Zonse zimene n'za Chauta wathu

Trad. Chichewa, Malawi

(Everything belongs to our God. All power, everything good.)

(Alles gehört Gott. Alle Macht, alles Gute.)

(Tout appartient à notre Dieu. Toute puissance, toute bonté.)

(Todo pertenece a nuestro Dios. Todo poder, todo lo bueno.)

Chauta means 'owner of the rainbow'.
Chauta bedeutet 'dem der Regenbogen zu eigen ist'.
Chauta signifie 'propriétaire de l'arc en ciel'.
Chauta significa 'propietario del arco iris'.

Copyright Acknowledgements

1 Music transcription © Maggie Hamilton
2 Words © Ulises Torres, translations © (P) Jaci Maraschin, (E) Fred Kaan, (G) tvd-Verlag Düsseldorf, (Fr) Marc Chambron
3 Translations © (E) Fred Kaan, (G) Strube Verlag, (Fr) Lutheran World Federation
4 Music transcription © Alexander Lingas
5 Translations © (E) Emmanuel Badejo, (G) Horst Bracks
6a Translations © (Fr) Réveil Publications, (Sp) Frederico J. Pagura
6b German text © Edwin Nievergelt; translations © (Fr) Marc Chambron, (Sp) Simei Monteiro
8 Translations © (E) Fred Kaan, (G) Dietrich Werner, (Fr) Marc Chambron, (Sp) Lutheran World Federation
9 Translations © (E) Fred Kaan, (G) Strube Verlag, (Fr) Marc Chambron
10 Translations © (G) Dietrich Werner, (Fr) Marc Chambron
11 Music arrangement © Oxford University Press; words © D. Sealy, translations © (E) Oxford University Press, (G) Dietrich Werner
12 Music arrangement © Simei Monteiro & Oxford University Press; translations © (E) Gerhard Cartford, (G) Strube Verlag, (Fr) Marc Chambron
13 Music and words © Wild Goose Resource Group, translations © (G) Thomas Laubach, (Fr) Hans Schmocker
14 Music © William A. Cross; words © GIA Publications, translation © Dietrich Werner
15 Music transcription © Maggie Hamilton
16 Music and words © Per Harling; music arrangement © Terry MacArthur; translations © (E) Per Harling, (G,Sp) World Council of Churches, (Fr) Joëlle Gouël

17 Music and words © Merete Wendler; translations © (E) Fred Kaan, (G) Dietrich Werner
18 Music arrangement © Rimantas Sliuzinskas; translations © (E) Per Harling, (G) Dietrich Werner
19 Music and words © Sr. Adelheid Wenzelmann; translations © (E) Fred Kaan, (Fr) Marc Chambron, (G) Lutheran World Federation
20 Music and words © Ateliers et Presses de Taizé; translation © (Fr) Marc Chambron
21 Translations © (E) Fred Kaan, (G) Dietrich Werner
22 Music and words © Pablo Sosa; translations © (E) Pablo Sosa, (G) Strube Verlag
23 Music © Norsk Musikforlag
24 Music and words © Harold Flammer Music, a division of Shawnee Press, Inc.; translations © (G) Dorothea Wulthorst, (Fr) Marc Chambron, (Sp) Lutheran World Federation
25 Words © 1968 Stainer & Bell; translations © (G) Strube Verlag, (Fr) Marc Chambron
26 Translations © (G) Strube Verlag, (Fr) World Council of Churches
27 Music © Ateliers et Presses de Taizé
29 Music © I-to Loh; words © Ruth Duck; translations © Lutheran World Federation
30 Music adaptation © Christian I. Tamaela
31 Music arrangement © Oxford University Press; words © Norman C. Habel; translations © (G) Dietrich Werner, (Fr) Marc Chambron
32 Music © Knut Nystedt; words © Sverre Therkelsen; translations © (E) Oxford University Press, (G) Sverre Therkelsen
33 Music and words © Olli Kortekangas; translations © (E) Oxford University Press, (G) Dietrich Werner, (Fr) Marc Chambron

34 Music and words © Willow Connection; instrumental descant © Oxford University Press; translations © (G) Dorothea Wulfhorst, (Sp) Lutheran World Federation

35 Music © tvd-Verlag; words © (G, F, Sp) Basler Mission; translation (E) © Oxford University Press

36 Music and words © World Alliance of Reformed Churches; translations © (E) Oxford University Press, (G) Dietrich Werner, (Fr) Marc Chambron

37 Music and additional words © GIA Publications; translations © (G) Lutheran World Federation, (Fr) Marc Chambron

38 Music and words © 1987 Bernadette Farrell (published by OCP Publications); translations © (G) Lutheran World Federation, (Fr) Marc Chambron

39 Music and words © Evangelikus Sajtoosztaly; translations © (E) Oxford University Press, (G) Strube Verlag

40 Music transcription © Maggie Hamilton

41 Music and words (H,E) © Charles Vas; translation (G) © Lutheran World Federation

43 Music © Dinah Reindorf

44 Music © Oxford University Press, translation © (E) Jeffrey T. Myers

45 Music (Dieter Trautwein arr. Hartmut Stiegler) and words © Strube Verlag; translations © (E) 1985 Stainer & Bell, (Fr) M. L. Munger and L. Levis

46 Music and words © Ateliers et Presses de Taizé

47 Music © Barrie Cabena; translations © (Fr) Marc Chambron, (G) Dietrich Werner

48 Music © Ateliers et Presses de Taizé

49 Music © Les Éditions du Cerf; translations © (E) Fred Kaan, (G) Dietrich Werner

50 Music © Evangelical Lutheran Church in America; translations © (G) Dietrich Werner, (Fr) Marc Chambron, (Sp) Lutheran World Federation

51 Music © Margaret Fleming; words © Frances Wheeler Davis; translation © Dietrich Werner

52 Music arrangement © Gerre Hancock; translations © (G) Strube Verlag, (Sp) Federico J. Pagura

53 Translation (Fr) © Centre National de Pastorale Liturgique

54 Words © Curso de Formação e Atualização Litúrgico Musical; translations © (E) Fred Kaan, (G) Ari Kuebelkamp, (Fr) Marc Chambron

55 Music © I-to Loh; words © Shirley Murray; translations © (G) World Council of Churches, (Fr) Joëlle Gouël

56 Music © Swee Hong Lim; words © Maria Ling; translations © (M) Dong Li, (G) Lutheran World Federation

57 Music and words © Oxford University Press; translations © (G,Sp) Lutheran World Federation, (Fr) Marc Chambron

58 Music © Zimbabwe Catholic Association of Sacred Music; translations © (E) Maggie Hamilton, (G) Lutheran World Federation

59 Music arrangement and words (including translations) © Vereinte Evangelische Mission

60 Music © þorkell Sigurbjörnsson; words © Sigurbjörn Einarsson; translations © (E) Per Harling, (G) Dietrich Werner

61 Music © T.Ilmari Haapalainen; words © Lars Thunberg; translations © (Fi) Kirkon Keskusrahasto, (E) Oxford University Press, (G) Jürgen Henkys

63 Music and words © Caroline Rockson; music transcription © Dinah Reindorf; translations © (E) Oxford University Press, (G,Fr,Sp) World Council of Churches

64 Music arrangement © Terry MacArthur and Oxford University Press; translations © (G) Horst Bracks, (Fr) Marc Chambron, (Sp) Lutheran World Federation

65 Music and words © Borealis Music; translations © (Fr) Marc Chambron, (G) Dietrich Werner

66 Music and words © Roger Courtney; arrangement © Oxford University Press; translations © (G) Lutheran World Federation, (Fr) Marc Chambron

67 Music and words © Shalom Community; arrangement © Oxford University Press; translations © (G) Lutheran World Federation, (Fr) Marc Chambron

68 Music and words © Bayiga Bayiga; translations © (E) Oxford University Press, (G) Dietrich Werner, (Sp) Lutheran World Federation

70 Music and words © Geonyong Lee; instrumental descant © Oxford University Press; translations © (E) Marion Pope, (G) Strube Verlag, (Fr) Robert Faerber, (Sp) Lutheran World Federation

71 Translations © (E) Alan Luff, (G) Lutheran World Federation, (Fr) Marc Chambron

72 Music © Homero Perera; arrangement © Terry MacArthur; words © Frederico J. Pagura; translations © (E) Len Lythgoe, (G) World Council of Churches

73 Music and words © GIA Publications; translations © (G) Dorothea Wulfhorst, (Fr) Marc Chambron

74 Words © Hope Publishing/CopyCare; translation © Dietrich Werner

75 Music © Ron Klusmeier; words © 1989 Stainer & Bell; translation © Dietrich Werner

76 Music arrangement © Maggie Hamilton; translations © (E) Francis Brienan and Maggie Hamilton, (G) Strube Verlag

77 Music arrangement © World Council of Churches; translations © (E) I-to Loh, (G,Fr) World Council of Churches

78 Music © Francisco F. Feliciano; words © Rolando S. Tinio; translations © (E) Rolando S. Tinio, (G) Dietrich Werner

79 Music and words © El Kasilik; translations © (E) John Campbell, (G,Sp) Lutheran World Federation, (Fr) Marc Chambron

80 Music (with arrangement by Raquel Mora Martinez) and words © GIA Publications; translations © (E) Choristers Guild, (G) Strube Verlag, (Fr) Marc Chambron

81 Music arrangement © Oxford University Press; translations © (E) Oxford University Press, (G) Dietrich Werner, (Fr) World Council of Churches

82 Music transcription © Maggie Hamilton

83 Music and words © Centre National de Pastorale Liturgique; translations © (E) 1974 Stainer & Bell, (G) Bärenreiter-Verlag

84 Music and words © Clara Ajo and Pedro Triana; music arrangement © Oxford University Press

85 Music and words © Mabel Wu; music arrangement © Y. Y. Wong; translations © (G) Lutheran World Federation, (Fr) Marc Chambron

86 Music © Francisco F. Feliciano; words © Rabindranath Tagore and (Seong-Won Park) World Alliance of Reformed Churches; translation © Reinhild Traitler

87 Translations © (G) Dietrich Werner, (Fr) Marc Chambron, (Sp) Lutheran World Federation

88 Music arrangement and English translation © General Board of Global Ministries; German translation © Dietrich Werner

89 Music and words © Universidad Biblico Latinoamericano; translations © (P) Simei Monteiro, (E) Fred Kaan, (G) Dorothea Wulfhorst

92 Music arrangement © Oxford University Press; words © General Board of Global Ministries; translation © Dietrich Werner

93a Music adaptation by Marcia Pruner and harmonization by John Campbell © GIA Publications; translation © Lutheran World Federation

93b,c Music adaptation by Richard Proulx and harmonization by John Campbell © GIA Publications; translation © Lutheran World Federation

94 Music and words © Teresita Savall; translations © (E) Fred Kaan, (G) Dorothea Wulfhorst, (Fr) Marc Chambron

95 Music arrangement © Oxford University Press; translation © Lutheran World Federation

96 Music and words © Rima Nasir Tarazi; translation © Dietrich Werner

97 Words and music © Simei Monteiro; music introduction © Oxford University Press; translations © (Sp) Jorge Rodríguez and Pablo Sosa, (E) World Council of Churches, (G) Strube Verlag

98 Music arrangement © Oxford University Press; translations © (E) Fred Kaan, (G) Lutheran World Federation

99 Music © Roland Forsberg; words © Anders Frostenson; translations © (E) Oxford University Press, (G) Jürgen Henkys

100 Music © Anfinn Øien; words © 1985 Stainer & Bell; translation © Dietrich Werner

101 Music and words (P,E) © General Board of Global Ministries; translations (G) Lutheran World Federation, (Sp) Pablo Sosa

102 Music and words © GIA Publications; translations © (E) Fred Kaan, (G) Dietrich Werner

103 Music and words © The Church of the Lord (Aladura); translations © World Council of Churches

104 Music © Francisco F. Feliciano; translations © (G) Strube Verlag, (Fr) Marc Chambron

105 Music arrangement © Oxford University Press; words © 1993 Stainer & Bell; translation © Dietrich Werner

106 Music © Bernard Smilde; words © Joke Ribbers; translations © (E) Fred Kaan, (G) Dietrich Werner, (F) Marc Chambron

107 Music and words © OCP Publications; translations © (G) Dorothea Wulfhorst, (Fr) Marc Chambron

108 Translation (G) © Lutheran World Federation

109 Music © Yusuf Khill; translation (G) © R. Renner and U. Schwendler

110 Translations © (E) Fred Kaan, (G) Weltgebetstagskommittee, Stein, (Sp) Lutheran World Federation

111 Music transcription © Maggie Hamilton; words adaptation © Fulata Moyo

Addresses of organisations owning copyrights reproduced in this book

Asian Institute for Liturgy and Music
PO Box 10533
Quezon City 1112
Philippines

Ateliers et Presses de Taizé
F-71250 Taizé-Communauté
France

Bärenreiter-Verlag
Heinrich-Schütz-Allee 35
D-34131 Kassel
Germany

Borealis Music
311-1424 Walnut Street
Vancouver, BC
V6J 3R3
Canada

Centre National de Pastorale Liturgique
4 Avenue Vavin
75006 Paris
France

Choristers Guild
2834 W Kingsley Road
Garland
TX 75041-2498
USA

The Church of the Lord (Aladura)
PO Box 71
Shagamu
Ogun State
Nigeria

CopyCare Limited
PO Box 77
Hailsham
East Sussex
BN27 3EF
United Kingdom

Curso de Formação e Atualização Litúrgico Musical
Rua Afonso de Freitas, 704
Paraíso
CEP 04006-052
São Paulo–SP
Brazil

Les Éditions du Cerf
29 bd La Tour-Maubourg
75340 Paris Cedex 07
France

Evangelical Lutheran Church in America
8765 W Higgins Road
Chicago
IL 60631
USA

Evangélikus Sajtoosztaly
c/o Evangelical Lutheran Church in Hungary
Szilagyi Erzsebet Fasor 24
1125 Budapest
Hungary

General Board of Global Ministries
GBGMusic
475 Riverside Drive
New York
NY 10115
USA

GIA Publications, Inc.
7404 South Mason Avenue
Chicago
IL 60638
USA

Kirkon Keskusrahasto
PO Box 185
FIN-00161 Helsinki
Finland

Lutheran World Federation
PO Box 2100
Route de Ferney 150
CH-1211 Geneva 2
Switzerland

Music Sales (Harold Flammer Inc.)
8–9 Frith Street
London
W1V 5TZ
United Kingdom

Norsk Musikforlag A/S
Postboks 1499 Vika
0116 Oslo
Norway

OCP Publications
5336 NE Hassalo
Portland
OR 97213
USA

Oxford University Press
Music Copyright Department
Great Clarendon Street
Oxford
OX2 6DP
United Kingdom

Réveil Publications
20 rue Calliet
BP 4464
F-69241 Lyon Cedex 04
France

Shalom Community Inc.
1504 Polk
Wichita Falls
Texas 76309
USA

Stainer & Bell Ltd
Victoria House
23 Gruneisen Road
London
N3 1DZ
United Kingdom

Strube Verlag
Pettenkoferstr. 24
D-80336 München
Germany

tvd-Verlag
Postfach 321111
D-40426 Düsseldorf
Germany

Universidad Biblico Latinoamericano
Calle 3
Avenidas 14 y 16 - Apdo. 901-1000
San José
Costa Rica

Vereinte Evangelische Mission
Rudolfstr. 137
D-42285 Wuppertal
Germany

Weltgebetstagskommittee, Stein
KlensVerlag GmbH
Postfach 320620
D-40421 Düsseldorf
Germany

Wild Goose Resource Group
Iona Community
Pearce Institute
840 Govan Road
Glasgow
G51 3UU
United Kingdom

Willow Connection Pty. Ltd
PO Box 341
Dee Why
NSW 2099
Australia

World Alliance of Reformed Churches
PO Box 2100
Route de Ferney 150
CH-1211 Geneva 2
Switzerland

World Council of Churches
PO Box 2100
Route de Ferney 150
CH-1211 Geneva 2
Switzerland

Zimbabwe Catholic Association of Sacred Music
PO Box CY 738
Causeway
Harare
Zimbabwe

Liturgical Index/Liturgischer Index/
Index liturgique/Índice litúrgico

This index contains suggestions of songs that might be used at particular points during the liturgy, reflecting a range of styles and traditions.

Dieser Index schlägt Lieder unterschiedlicher Stilrichtungen und Traditionen für verschiedene liturgische Zwecke vor.

Cet index suggère des cantiques, de styles et de traditions variés, qui peuvent être chantés aux différents moments de la liturgie.

Este índice contiene sugerencias de cantos de diversos estilos y tradiciones para ser usados en los diferentes momentos de la liturgia.

Acclamation/Akklamation/Acclamation/Aclamación

A ivangeli ya famba (1)	Mozambique
A Palavra do Senhor (3)	South America
E toru nga mea (21)	Aotearoa New Zealand
Praise the God of all creation (73)	USA
Rakanaka Vhangeri (76)	Zimbabwe
Siyabonga (82)	Southern/East Africa

Affirmation of faith/Glaubensbekenntnis/Affirmation de la foi/Afirmación de fe

Jikelele (40)	South Africa
Reamo leboga (77)	Botswana
We believe, Maranatha (104)	Philippines
Wij geloven één voor één (106)	Netherlands
Zonse zimene n'za Chauta wathu (111)	Malawi

Agnus Dei: Lamb of God/Lamm Gottes/Agneau de Dieu/Cordero de Dios

Jesus, Lamb of God (37)	USA
Lamb of God (47)	Canada
Shang di de gao yang (85)	China
Vos sos el destazado en la cruz (102)	Nicaragua
Ya hamalaLah (109)	Palestine/Israel

Alleluia/Halleluja/Alléluia/Aleluya

Aleluya Y'in Oluwa (5)	Nigeria
Alleluia (7)	Ukraine
Haleluya Puji Tuhan (30)	Indonesia
Iloitse, maa! (33)	Finland

Amen/Amen/Amen/Amén

The Great Thanksgiving (Amen) (93*c*)	USA

Baptism/Taufe/Baptême/Bautismo

Crashing waters at creation (14)	Canada
O healing river (64)	African American
O living water, refresh my soul (67)	USA

Blessing/Segen/Bénédiction/Bendición

Bwana awabariki (10)	Tanzania
Komm, Herr, segne uns (45)	Germany
May the love of the Lord (56)	Singapore
Tú eres amor (94)	Argentina
Yarabba ssalami (110)	Palestine

Communion/Abendmahl/Communion/Comunión

Bendice, Señor, nuestro pan (9)	Argentina
Du bist unser alles (19)	Germany
Eat this bread (20)	France
El cielo canta alegría (22)	Argentina
Jesus, Lamb of God (37)	USA
Jesus, Lamb of God, bearer of our sin (38)	England
Let us break bread together (52)	African American
Loving Spirit (55)	Taiwan
Sa harap ng pagkaing dulot mo (78)	Philippines

Confession of sins/Sündenbekenntnis/Confession du péché/Confesíon de pecados

A ti, Señor, te pedimos (2)	Chile

Daily prayer/Stundengebet/Prière quotidienne/Oración diaria

asato maa sad gamaya (8)	India
Kiedy ranne wstają zorze (44)	Poland
La ténèbre (46)	France
Le Seigneur fit pour moi des merveilles (49)	France
Let my prayer rise before you (50)	USA
Nú hverfur sól í haf (60)	Iceland
Nu sjunker bullret (61)	Finland
Vai com Deus! (101)	Brazil
We believe, Maranatha (104)	Philippines

Eucharistic prayer/Abendmahlsgebet/Prière eucharistique/Oración eucarística

The Great Thanksgiving (Christ has died) (93*b*)	USA

Gathering/Eröffnung/Rassemblement/Al congregarse
All people that on earth do dwell (6a)
Come, Holy Spirit, descend on us (13)   Scotland

Gloria: Glory to God/Ehre sei Gott/Gloire à Dieu/Gloria a Dios
Gloria a Dios (26)   Peru
Gloria in excelsis Deo! (27)   France
Slava voveki Tsaryu (88)   Russia
Utukufu (98)   Kenya

Invocation/Anrufung/Invocation/Invocación
Come, Holy Spirit, descend on us (13)   Scotland
Dvasia Šventoji, ateiki (18)   Lithuania
Nyame ne Sunsum, sian brao! (63)   Ghana
O God, we call (65)   Canada
O Saint Esprit, viens sur nous (68)   Cameroon
Ososǒ (70)   Korea
Wa Emimimo (103)   Nigeria

Kyrie: Lord, have mercy/Herr, erbarme dich/Seigneur, aie pitié/Señor, ten piedad
Gospodi pomilui (28)   Ukraine
Khudaya, rahem kar (42)   Pakistan
Kyrie eleison (43)   Ghana
Oré poriajú verekó (69)   Paraguay
Señor, ten piedad de nosotros (84)   Cuba
Xin Chua thuong xot chung con (108)   Vietnam

Litany/Litanei/Litanie/Letanía
Word of justice (Litany of the Word) (107)   England

Offertory/Gabenbereitung/Offertoire/Ofertorio
Te ofrecemos nuestros dones (89)   Argentina

Our Father/Vaterunser/Notre Père/Padrenuestro
Fader vår (23)   Norway

Post-communion prayer/Dankgebet/Post-communion/Oración de Acción de Gracias
Strengthen for service, Lord (87)

Psalmody/Psalmodie/Psalmodie/Salmodia
All people that on earth do dwell (Psalm 100) (6a)
Cantai ao Senhor (Psalm 98) (12)   Brazil
Iloitse, maa! (Psalm 100) (33)   Finland

Laudate omnes gentes (Psalm 117) (48)	France
Let my prayer rise before you (50)	USA
The deer longs for pure flowing streams (Psalm 42) (92)	Caribbean

Response/Antwortgesang/Réponse/Responsorio

For God alone (24)	Canada
I will live for you alone (34)	Australia
Laudate omnes gentes (48)	France
Le Seigneur fit pour moi des merveilles (49)	France
Let my prayer rise before you (50)	USA
O God, we call (65)	Canada
Ouve, Senhor, eu estou clamando (71)	Brazil
Salwalqulubi (79)	Lebanon
We believe, Maranatha (104)	Philippines

Sanctus: Holy/Heilig/Saint/Santo

Du är helig (16)	Sweden
Musande Mambo Mwari (58)	Zimbabwe
Santo es nuestro Dios (80)	El Salvador
Santo, santo, santo (81)	Argentina
The Great Thanksgiving (Holy, holy, holy Lord) (93*a*)	USA

Sending/Sendung/Envoi/Envío

Tembea na Yesu (90)	Kenya
Those who wait upon the Lord (95)	North America
Thuma mina (91)	South Africa
Truth is our call (96)	Palestine
Vai com Deus! (101)	Brazil
We shall go out with hope of resurrection (105)	Northern Ireland

Table grace/Tischgebet/Prière pour le repas/Bendición de los alimentos

Be present at our table, Lord (Table graces) (6*b*)	
Bendice, Señor, nuestro pan (9)	Argentina
Du bist unser alles (19)	Germany
Sa harap ng pagkaing dulot mo (78)	Philippines

The Peace/Friedensgruss/La paix/La paz

| Put peace into each other's hands (75) | Canada |

Trisagion: Holy God/Dreimalheilig (Heiliger Gott)/Dieu saint/Santo Dios

| Agios o Theos (4) | Greece |

Biblical References/Übersicht der Bibelstellen/ Références bibliques/Referencias bíblicas

First Lines and Titles/Liedanfänge und Titel/ Premières lignes et titres/Primeras líneas y títulos

Titles are shown in *italics*/Die Titel sind *kursiv* gedruckt/
Les titres apparaissent en *italiques*/Los títulos aparecen en *cursiva*

A ivangeli ya famba	1
A notre table, viens, Seigneur	6*b*
A nuestra mesa ven, Señor	6*b*
A Palavra do Senhor não lhe voltará vazia	3
A ti, Senhor, te pedimos	2
A ti, Señor, te pedimos	2
Agios o Theos	4
Agneau de Dieu	47
Agnus Dei	37
Aleluya Y'in Oluwa	5
All people that on earth do dwell	6*a*
All the earth, praise!	33
Alle Schmerzen trugst du selbst am Kreuz	102
Alleluia	7
Alleluia, praise the Lord	5
Allons semer les graines de paix	66
Alma, bendice al Señor, Dios potente de gloria	53
Amen	93*c*
As evening falls, Lord, we lay our burdens	61
Asante Yesu, amin	82
asato maa sad gamaya	8
Auf Gott allein harrt meine Seele, stille	24
Be present at our table, Lord	6*b*
Because he came into our world and story	72
Bendice, Señor, nuestro pan	9
Bí, a Íosa, im chroíse i gcuimhne gach uair	11
Bless and keep us, Lord, in your love united	45
Bwana awabariki	10
Cada lugar de este mundo	35
Cantad al Señor un cántico nuevo	12
Cantai ao Senhor um cântico novo	12
C'est Dieu seul que mon âme attend en silence	24
C'est la foi de tout chrétien	106
Chantons au Seigneur un nouveau cantique	12
Chaque lieu sur cette terre	35